A Flash of Light
The Science of Light and Colour

CW00524118

A Flash of Light
The Science of Light and Colour

Edited by

Mark Lorch
University of Hull, UK
Email: m.lorch@hull.ac.uk

Andy Miah
University of Salford, UK
Email: a.miah@salford.ac.uk

ROYAL SOCIETY
OF **CHEMISTRY**

Print ISBN: 978-1-78262-731-9
EPUB eISBN: 978-1-78262-859-0

A catalogue record for this book is available from the British Library

© The Royal Society of Chemistry 2016

Reprinted 2020

Published by The Royal Society of Chemistry,
Thomas Graham House, Science Park, Milton Road,
Cambridge CB4 0WF, UK

Registered Charity Number 207890

Visit our website at www.rsc.org/books

Printed in the United Kingdom by CPI Group (UK) Ltd, Croydon, CR0 4YY, UK

Preface

A Flash of Light comes at a radical time in the history of scholarly publishing. With mobile and digital books capturing more of the attention of readers, the number of published scholarly articles doubling every decade, and a growing need to reimagine the book for the 21st century, our book is a product of these times.

Typically, when a scientist has the initial spark of an idea, it might be years before the fruits of their labour is read. In between, grant proposals are written – and hopefully won – researchers are appointed to help carry out the work, papers are eventually written, peer reviewed, and finally, after what can be 5 years in total, these findings are published and have the chance of reaching the general population. Yet, even here, more work is needed by the publisher to ensure a wider audience and, typically, academics must take their work to intermediary platforms, such as the media, or book fairs to reach beyond their peers.

The duration of this process, coupled with questions about the integrity of the peer review system, have led some academics to interrogate and propose new working models for researchers and, perhaps since the digital age, academics have found outlets for their work to quench a growing desire to reach a wider public. In recent times, platforms like The Guardian's science website, the Huffington Post and more, recently, the Conversation, have become spaces in which academics can write differently and reach new audiences.

A Flash of Light: The Science of Light and Colour
Edited by Mark Lorch and Andy Miah
© The Royal Society of Chemistry 2016
Published by the Royal Society of Chemistry, www.rsc.org

At the same time, the rise of e-readers and e-publishing more widely provide greater opportunities to get ideas out fast. This was the pre-text for *A Flash of Light*, which aimed to turn the academic publishing model on its head and bring together some gifted writers and thinkers to fly in the face of established practices. The working hypothesis was that, if you could get a number of authors together in the same room for 2 days working intensively and without breaks or distractions from all of the other things that academic life brings, we could produce an amount of work equivalent to that which would otherwise take a year or two to accomplish.

We found the perfect environment in which to run this experiment within the Manchester Science Festival, through the University of Salford's partnership as principal educational sponsor. In this year, the university sought to produce work that narrowed the gap between publication and presentation – to allow scientists not only to present their findings within a science festival, but to make discoveries that would lead to published work.

To assist this hyper-short book sprint, we also partnered with an organization whose values resonated with that of the project, namely, The Conversation and a starting point for discovering authors was to identify writers for their platform who might be suitable. Working closely with the science editors at The Conversation, Stephen Harris and Miriam Frankel, a dream team of writers was identified and brought together. Consequently, nine academic authors, two professional science writers, four graphic design illustrators, two copy editors, and a facilitator were brought together, to make it all happen.

The result of this frantic weekend was 9 chapters comprised of around 30 000 words, supplemented by around 20 illustrations. Those chapters were messy, still needed editing, referencing and some tidying up, but they were good. They had a sense of pace and energy and they hung together into a fascinating story covering an incredible range of light related topics.

Our book takes an epic journey, which starts by exploring the colours of the universe and the sky above our heads. It covers light you never knew you could see and how light influenced the evolution of animals, we examine the psychology of colour and vision before looking at how we have harnessed light for our own gains.

We learnt a fabulous amount in our weekend sitting around a table frantically researching and typing. Some fascinating material has not made it into the main text, but is worth mentioning. For example, we spent an hour or two brainstorming the topic of the book and, whilst we pretty much ended up writing what we wanted, we all got very excited about where colour is actually located. Discussions ventured from colour blindness, to the experiences of people who have had their sight restored and synesthesia. In the course of their discussion our facilitator, Mark Cutter, noted that he is a governor of the Royal Institute for the Blind, and, 10 minutes later, he had Denise Leigh, a blind opera singer with synesthesia, on the phone talking to us. Her condition means that she can see sounds and she described the incredible ribbons of colour she sees whilst singing, the hues of her children and the blessing of her being a synesthete. Denise's story exemplifies the brief and the rapid journey we went through during the course of the weekend, where the group sat around the table for 22 hours throwing stories, facts and figures at each other.

More often than not, edited books in academia are made without ever the authors coming together to work on a common core manuscript and this experiment sought to transform this model. However, it was not just an exercise in productivity and work flows. It was also an inquiry into how one makes the act of writing a performance and how this ritual of real-time collaboration can create a sense of history that can enrich our lives. Time will tell how our individual authors feel about the work they produced and the publication that resulted, but at the very least, we have shown that a lot more can get done, a lot quicker, by aggregating knowledge and focusing its discovery down in a very short amount of time.

Crucially, the book would not have happened without the additional support and belief in us by the Royal Society of Chemistry, particularly the hard work of Cara Sutton. We are tremendously grateful for the Royal Society of Chemistry's investment and willingness to try something completely unprecedented. Here again, we feel that this relationship was atypical, where the publisher had a closer intellectual involvement with the generation of our words than is often the case.

We'd also like to thank Ian Morris, Heather Holst and Liz Bryan for their wonderfully amusing and informative illustrations.

And finally thanks to the University of Hull and University of Salford for providing financial support for the project (we had to eat and sleep!) and particularly Salford for giving us a place to write.

The authors, editors and illustrators, and the conversation that lead to *A Flash of Light*. Image credit: Ian Morris.

Contents

A Flash of Light: The Science of Light and Colour
Edited by Mark Lorch and Andy Miah
© The Royal Society of Chemistry 2016
Published by the Royal Society of Chemistry, www.rsc.org

CHAPTER 1

Where Is Colour?

Giuliana Mazzoni

University of Hull, UK

> *You might expect a book about light to start with some physical description of electromagnetic waves and how their properties relate to the colours that we see. Well we will get to that later, but we're going to start somewhere else entirely; the place where colour actually exists—in your mind ... And we're going to explore that place with some near future science fiction about a gene therapy that could soon be available to treat colour blindness (or more accurately colour deficiency).*

At the moment, the therapy has only been fully tested on animals, but the results of initial clinical trials in humans suggest that it might be extended to pathological conditions in human colour vision in the relatively near future.[1] Let's imagine what it would be like for a person suffering from colour blindness who undergoes this gene therapy and recovers colour vision.[†]

[†]The experience of colour blindness and recovering colour vision is based on scientific data currently available (although recovery does not as of yet depend on gene therapy).

A Flash of Light: The Science of Light and Colour
Edited by Mark Lorch and Andy Miah
© The Royal Society of Chemistry 2016
Published by the Royal Society of Chemistry, www.rsc.org

1.1   A SUCCESSFUL STORY FROM THE FUTURE

"As I was walking to the train station this morning, I noticed the colours of the leaves on the deciduous trees around the house—gold, bright yellow, pale green, deep red. Beautiful. But, 'where is colour?' I was asking myself. The question comes from my own experience as someone who used to be colour blind. That's right, colour blindness can be corrected. In the beginning, my eyes did not work properly, but after a new rather revolutionary surgical intervention, I could see colours as they appear in the external world to other members of my species. What an experience!

My form of colour blindness is called achromatopsia and it is rather rare. You see, it turns out that most colour blind people do actually see some colour. What's more, there are various forms of colour blindness; some can see a couple of colours, although somewhat dull, others can see much less colour than most. Fancy names are associated with these various forms of pathology. For example, dichromats (mostly men) can see only two ranges of colour and, among those, deuteranopes present problems discriminating between red and green while others with protoanopia cannot see red and similar hues, those with tritanopia fail to discriminate blue hues. Other people can only see one colour, monochromats. Finally, there are people like me, who have complete color blindness, otherwise known as achromatopsia. People of my (original) disposition are completely colour blind, like the inhabitants of the remote island (the Micronesian atolls of Pingelap and Pohnpei) described by Oliver Sacks.[2] Apparently, they also see the world the way I saw it, in shades of grey.[3] I was the one in 40 000 born that year with achrmatopsia: 39 999 with good vision, and me. But all that has changed now, I can finally see colour.

Yet, a question occurred to me in my current state: Now that I see the colour of the leaves, does it mean that leaves really have colour? Which was the version of the world closest to the truth: is the world coloured? Or is colour just in the eye and the brain of the beholder? Before surgery, my world was mostly monochromatic. It is a rather strange experience to recover colour vision, especially at the beginning, when colour patches do not match the shape of the objects to which they are supposed to belong.[4–6] I could see the shape of my car, but the red was floating somewhere near to it, and at first it did not 'stay' within the car shape. It was when I was recovering, and seeing those coloured patches outside of the objects they were supposed to belong to, that I wondered if colours do indeed exist in the external world, or whether they are only 'in the mind of the beholder'.[7]

In some cases achromatopsia is determined by cerebral damage to areas in the occipital lobes, mostly an area called V4 (for visual) and specifically the lingual and fusiform gyri. Others, like mine are a special condition, in which little cells in the retina—called cones—were not working, while the brain was still intact.[8,9] While most people think about the retina as simply a part of the eye, it is actually a part of the brain that was placed at the bottom of the eye. When all is going well, cones convert the electromagnetic waves of light sensory stimulus into electrical neural impulses.

Going back to my pathology, my cones were not working, because they were missing crucial proteins—opsins. These proteins react to specific light wave lengths and make colour visible. Without these opsins within the cones, there is no colour vision.[10]

During surgery, genes that coded for the opsin proteins were artificially inserted into my cones giving my vision the range of colour that most people enjoy, more or less. The surgery was rather scary, but it had been successfully tested on our genetic cousins—other primates—before me, and had undergone some initial clinical trials on humans, so I was hopeful. It is a form of gene therapy and I am certain many more people in a similar condition as mine will volunteer to have it done. I will talk about the surgery later, but the core of my question remains, where is colour? Is it in the objects that we see, or is it in our eyes, or in the brain?

Living with colour blindness is not easy. A person who can see in colour cannot understand enough what colour brings. For example, there are emotional reactions innately connected to colour. Males seem to respond better to blue, females to red.[11] Now, just to be clear, this doesn't explain why blue is sometimes associated with boys and why girls are associated with pink! That's a whole different ball game in sociological research. Nevertheless, even some forms of therapy are based on the realisation that different colours elicit different emotions in different people, even if some research suggests that specific emotions are elicited by specific colours in everybody (e.g. orange elicits friendliness, red excitement, green peace, etc). Also, people seem to respond to colour when making a purchase *(does it mean advertisement and marketing strategies had no effect on me?) and colour responses can often play a big part in how our commercial world is seen. Curiously, the reason why Facebook's design is mostly blue is not because it was made by boys, but because its founder Mark Zuckerberg is red-green colourblind and can see blue best. Facebook apart, colour affects people's purchasing decisions, and marketing companies base brand colours on existing research in order to endear people to their products. Research made by companies has shown that specific colours are associated with specific attributions of product characteristics. For example, black is found to be associated with the idea of stability, credibility, strength, power, professionalism, accuracy; green with natural, organic, education, adventurous, calming; blue with cleanliness, medical, professional, judicial, business-like. Hence, to substantially simplify matters, black is good for solid corporations, green for ecological business and tourism, blue for medicine and science, etc. I looked at the website, https://blog.kissmetrics.com/color-psychology and indeed label colours convey—even to me—different impressions and emotions.*

Typically, we also respond differently to pills of different colour. For example, even if they are totally inert placebos, i.e. basically just sugar pills, blue pills have a relaxing, calming effect, as do anxiolitics.

Red pills, on the contrary, have a stimulant effect, which is similar to caffeine. Taking a coloured pill has a greater effect than taking the same pill but colourless, meanwhile popping a coloured pill twice has an even stronger effect than taking it once. So, with all the perks that come with colour vision, being able to see only dark and light and shades of grey makes life rather difficult. Furthermore, although I have not been tested then, I might not have shown the same purchasing behaviors or the same placebo effect of red vs blue pills as most people do.

In some cases, colour is also essential for identifying objects. Basic object identification is due to the ability to detect the contour, the outline, of the shape of the object. In real life, the contour is not a black line, it is most of the times a difference in colour, difference linked to luminance, colour intensity, colour contrast. So, a person who is colour blind can experience major difficulties in identifying objects. While there are many different cues one can use that are not based on colour to distinguish objects, perception can remain partly impaired.[12] Fortunately for me, I could see depth and so I never fell down any precipices, but driving was not really possible.

So, when I was colour blind, I could still see shapes and objects, distinguish objects from the context and the background, but lack of colours made object recognition difficult. The fact that colours are important is also evidenced by the history of medical doctors and scientists (chemists, for example) who reported their specific impairment and the negative effect of colour blindness on their professional activity,[13] and the need to take remediational steps to overcome such limitations. Even if they are not painters, their professional life can be impaired.

Now they have tested me using a relatively easy method that they call nonmetric multidimensional scaling. It consists of judging (overtly) the similarity of colour hues. I am not performing yet as people who always had colour vision. All three types of cones in my retina are working; short-wave (S), medium-wave (M) and long-wave (L) cones. I can now see magenta, cyan and green, although the mapping of S, M and L cones on the three colours is far from perfect, though this is true for everybody.

When I was still recovering I was confused by colours for a while seeing, among other things, reddish frogs and bluish leaves! In normal people, green (medium) light waves activate the other two colours, as it is not possible to stimulate only M cones (so-called "green" cones). The other two are also stimulated to a certain degree, but the system has learned to discriminate among them. I needed to learn that too and the reddish frogs and bluish leaves became slowly greener over time. The opponent process of light vs. dark I always had been able to see. I am also starting to have the experience of blue vs. yellow and red vs. green.

I have talked a lot about colour, but again, still have not discovered its location.

When I said that I could not see colour, it was exactly like that, no colour coming from the outside world at all. However, there are people who, in spite of being unable to see colour at all with their eyes, still can see colour in their mind's eye. A great example comes from people who have synesthesia. These are somewhat special people in whom the five senses (and especially hearing and vision, but also touch and smell) are connected in an unusual way. For example, they can see musical notes, and music becomes a stream of visual elements. Often the visual association comes as colour.

For example I recently heard the case of a blind opera singer who sees the music she sings as a continuous stream of colours.[‡] Her children also evoke visual colours for her. Her daughter is very yellow, and her son is acid green. She cannot see any colour through her eyes, but she can see colour in her mind. It is true that she was able to see colours when she was young, before her vision completely deteriorated, but now her colour vision seems to be completely mediated by her brain. It seems that several non-blind musicians also see music in colour: Liszt, Rimsky-Korsakov, Sibelius, and the less known Joachim Raff who saw a different colour for each musical instrument.

So, once again, where is colour? Is it in the objects, as wavelengths, or in the brain as the experience of synesthesia suggests? The role of the brain in making sense of colour becomes even clearer in some visually impaired synesthetes. Such people can see colours they have never experienced in their life. One patient described by Vilayanur Ramachandran and Edward Hubbard[14] cannot see at all certain hues, because of a deficit in color receptors. However, when looking at numbers, his synesthesia enables him to experience colors in his mind that he has never seen in the real world. He calls these "Martian colors". Well, I wonder what colour are these Martian colours, and I wonder if one can discover what they look like by making someone with this capacity choose among the more than 800 Munsell cards that contain an incredibly large variation of hues.

The fact that, in other conditions, people can see colours that do not exist tells us that colour is not in the objects. Some people, for example, can 'decide' to see colour in patterns that are just shades of grey. But how does it happen? Research carried out in 2014 in the Department of Psychology at the University of Hull[15,16] examined a special skill present in about 10–15% of the population—that of hallucinating colours at will. People with this ability can see colour where no colour is presented. Such people are not at all gullible; they possess the enviable skill of being able to modify their perception as they like. They can stop experiencing pain, they can hallucinate sounds and even figures at will. So these extraordinary people were presented with Mondrian-style coloured patterns, or with the same

---

[‡]We (the authors of this book) interviewed Denise Leigh on the first morning of writing this book. Here descriptions of synthesia are told here.

Mondrian pattern, but this time in shades of grey. To a certain extent, the grey pattern resembled my kind of colour blindness.

When grey Mondrian patterns were shown, participants were invited to use their ability to alter their own colour perception and see a coloured pattern where there was none. They were then asked to identify on Munsell cards the exact colour hues they were seeing. In an additional condition, visual aftereffects of these intentional colour hallucinations were studied. For example, if you see red for a long time, for the rules of opponent process, you will see green if you then move your eyes to a white board. A substantial percentage of these particularly skilled individuals saw the grey Mondrian pattern in colour, with colours varying from red to orange to yellow to green. They never saw blue or purple or violet. Many of them also had visual aftereffects, which confirmed that they were not confabulating, nor being complacent with the requests of the experimenter. Rather, they were having a visual effect that involved colour when no colour was actually there.

Where do such 'hallucinated' colours come from? fMRI results on these individuals show that the same posterior brain areas that are involved in seeing *colour were involved when they hallucinated these colours.*[16] *Colour hallucinations are thus the result of the activation of one of the posterior areas of the brain, the lingual gyrus, which are part of the brain network responsible for seeing colour in the real world.*

So, going back to my question, where is colour? In some instances, specific light waves trigger retinal cells, that much is clear. However, colour vision can be achieved completely independently of the physical external stimulus and so this is not a sufficient explanation. Moreover, the brain produce colours, even if no light triggers any reaction inside the retina's cones.

What can I then conclude? Are leaves actually green? One must conclude that it is actually the activation of the visual areas responsible for colour vision that are ultimately responsible for colour vision. If microelectrodes are inserted in those colour areas, then the person sees colour, even with closed eyes. The reason why the brain can produce colour is that, strangely, the brain areas have functional specialization. As there are neuronal clusters that react specifically to simple features as horizontal lines, or vertical lines, and there are neural clusters that are specialized to see faces,[17] *there are also brain areas that, if activated, produce the experience of colour. In short, we are born to see colour. More broadly, the act of seeing is, more often than we think, actually an illusion. This does not imply that all we see is always a hallucination and that the physical stimuli never play a role. Quite to the contrary. Although there are many instances in which seeing is possible in the absence of external stimulation (e.g., visual hallucinations), the visual system is also constantly activated in response to external light. However, even when responding to light, one should not underestimate the immense difference between the nature of the external stimulus and the final nature of the percept, which is exclusively the result of neural and brain processes. The green of the leaves is inside me, it is in my brain.*

1.2   BACK TO THE PRESENT

While the previous fictional story illustrated the experience of a future person with complete achromatopsia who undergoes genetic surgery, here is briefly reported the case of an existing young man, Ethan, who presents a more common form of colour blindness. His vision is not in shades of grey, he can see some colours, but they are more dull than the way they normally look when the eye is intact. He can see some pink and green but, at times, the pink looks silver or blue, and green looks brown or yellow. This is due to how his retinal cones react to light waves of different length, which is different from the way cones normally work. According to his own account, his S cones, those that react to short wave length (blue range), seem to be relatively intact, which implies that he should be able to separate short waves from the others and see blue. He sees yellow because yellow is the opposite of blue. When activated, S cones cause us to see blue and, when inhibited, lead to seeing yellow. However, the medium (green) and long (red) cones do not discriminate well between these two different ranges of wave length, and the perceived colours then get muddled. So, Ethan sees green as brown. Recently, a friend gave him a very special pair of glasses, which separate medium from long light waves. To his utter astonishment, these glasses made it possible for him to see this difference and, for the first time, to see colours as they appear to an intact eye. Or, at least this is what he reports.

While these glasses are still an assistive device and not a cure, there is hope that, in the future, some types of colour blindness can be cured also in humans. Gene therapy has proved rather successful in a range of animals, including primates. Gene therapy is a very delicate procedure that involves the addition of the photo-sensitive pigments that are not working (the opsins that are missing in colour blind people) *via* a viral vector that is inserted in the appropriate cones. Red–green colour blindness, which is the most common form of partial colour blindness, derives from a deficit in either L or M cones, as it happens in Ethan's case. More specifically, the deficit is in the L and/or M opsins, the proteins that function as photopigments responsible for responding and elaborating long and/or middle light wavelengths. This is typically a genetic disorder. These people have dichromatic (two-colour) rather than the normal trichromatic (three-colour) vision. Adding the missing photopigment restores trichromatic vision. It was previously believed that no improvement in colour vision could be obtained unless the intervention was done on the very young (earlier sensory deprivation studies had shown that neural connections established during development would not appropriately process an input that was not present from birth). However, a 2009 study has demonstrated that gene therapy can be successful also in adult monkeys,[18] which gives hope for future similar interventions in adult humans. Trials on humans are already under way.[19]

Yet there are some problems in the delivery of the viral vector to the retina, as the virus has to be injected directly inside the retina using a needle. Besides being unpleasant, such a procedure carries the risk of infection. Furthermore, it is not yet known how many iterations of the injection one needs to go through in order to obtain a stable effect, and repeated injections can produce an immune reaction to the virus. So, it may well be a while before this technique is available to large numbers of people. But still, there is hope for a 'coloured future'.

REFERENCES

1. V. Sundaram, C. Wilde, J. Aboshiha, J. Cowing, C. Han, C. S. Langlo, R. Chana, A. E. Davidson, P. I. Sergouniotis, J. W. Bainbridge, R. R. Ali, A. Dubra, G. Rubin, A. R. Webster, A. T. Moore, M. Nardini, J. Carroll and M. Michaelides,

Retinal structure and function in achromatopsia: implications for gene therapy, *Ophthalmology*, 2014, **121**(1), 234–245.

2. O. Sacks, *The Island of the ColorBlind*, Knopf, New York, 1997.
3. J. A. Brody, I. Hussels, E. Brink and J. Torres, Hereditary blindness among Pingelapese people of Eastern Caroline Islands, *Lancet*, 1907, **1**(7659), 1253–1257.
4. C. J. Lueck, S. Zeki, K. J. Friston, M. P. Deiber, P. Cope, V. J. Cunningham, A. A. Lammertsma, C. Kennard and R. S. J. Frackowiak, The colour centre in the cerebral cortex of man, *Nature*, 1989, **340**, 386–389.
5. D. J. McKeefry and S. Zeki, The position and topography of the human colour centre as revealed by functional magnetic resonance imaging, *Brain*, 1997, **120**, 2229–2242.
6. R. Shapley and M. J. Hawken, Color in the cortex: single- and double-opponent cells, *Vision Res.*, 2011, **51**, 701–717.
7. R. N. Shepard and L. A. Cooper, Representation of Colors in the Blind, Color-Blind, and Normally Sighted, *Psychol. Sci.*, 1992, **3**(2), 97–104.
8. S. Kohl, T. Marx, I. Giddings *et al.*, Total colourblindness is caused by mutations in the gene encoding the alpha-subunit of the cone photoreceptor cGMP-gated cation channel, *Nat. Genet.*, 1998, **19**(3), 257–259.
9. B. Wissinger, D. Gamer, H. Jägle *et al.*, CNGA3 mutations in hereditary cone photoreceptor disorders, *Am. J. Hum. Genet.*, 2001, **69**(4), 722–737.
10. A. Ödeen and O. Håstad, Complex Distribution of Avian Color Vision Systems Revealed by Sequencing the SWS1 Opsin from Total DNA, *Mol. Biol. Evol.*, 2003, **20**(6), 855–861.
11. L. Ellis and C. Ficek, Color preferences according to gender and sexual orientation, *Pers. Individ. Differ.*, 2001, **31**(8), 1375–1379.
12. T. Gevers and A. W. M. Smeulders, Color-based object recognition, *Pattern Recognit.*, 1999, **32**, 453–464.
13. J. A. Spalding, Confessions of a colour blind physician, *Clin. Exp. Optom.*, 2004, **87**(4–5), 344–349.
14. V. S. Ramachandran and E. M. Hubbard, The phenomenology of synaesthesia, *J. Conscious. Stud.*, 2003, **10**(8), 49–57.

15. G. Mazzoni, E. Rotriquenz, C. Carvalho, M. Vannucci, K. Roberts and I. Kirsch, Suggested Visual Hallucinations In and Out of Hypnosis, *Conscious. Cogn.*, 2009, **18**, 494–499.
16. McGeown, G. Mazzoni, Venneri and Kirsch, Suggested visul hallucinations without hypnosis enhance activation in visual areas of the brain, *Conscious. Cogn.*, 2011, **21**(1), 100–116.
17. E. T. Rolls, Neurons in the cortex of the temporal lobe and in the amygdala of the monkey with responses selective for faces, *Hum. Neurobiol.*, 1984, **3**, 209–222.
18. K. Mancuso, W. W. Hauswirth, Q. Li, T. B. Connor, J. A. Kuchenbecker, C. Matthew, M. C. Mauck, J. Neitz and M. Neitz, Gene therapy for red–green colour blindness in adult primates, *Nature*, 2009, **461**, 784–787.
19. A. V. Cideciyan, W. W. Hauswirth, T. S. Aleman, S. Kaushal, S. B. Schwartz, S. L. Boye, E. A. M. Windsor, T. J. Conlon, A. Sumaroka, A. J. Roman, B. J. Byrne and S. G. Jacobson, Vision 1 year after gene therapy for Leber's congenital amaurosis, *N. Engl. J. Med.*, 2009, **361**, 725–727.

CHAPTER 2

What Are We Really Seeing?

Akshat Rathi

Quartz, London, UK

Directly and indirectly, humans interact with the light around them throughout their lives. We take the presence of light in our lives for granted. Yet, those who have never seen light, or who have lost the ability to see it, have tales to tell that can change our perspective of life. Their stories make us stand back and marvel at the magical ability that our eyes endow on us. In many ways, our lives become defined by this interaction of light particles with matter around them. So what happens when people lose the ability to see light? How does this change their experience of the world?

2.1 TWO WORLDS

Few blind people can tell their story better than John Hull, the author of the 1990 book *Touching the Rock*. Hull was born with a congenital condition that caused the development of cataracts in his early teens. The world grew darker and darker, until in 1980, at the age of 45, Hull totally lost his vision. "When I lost my sight,

A Flash of Light: The Science of Light and Colour
Edited by Mark Lorch and Andy Miah
© The Royal Society of Chemistry 2016
Published by the Royal Society of Chemistry, www.rsc.org

I suffered a lot from boredom. I just didn't know what to think about...That was a problem for me at first," he told Radiolab.[1]

Once when a friend commented that he felt lucky, in some way, that he will only remember his wife as young and beautiful, Hull felt that he was living a lie. The reality was that he had lost one of his senses and, thus, the world as he experienced it was only the summation of what his other senses could "show" him. His next decision would correct this "mistake": he decided to live a life without pictures at all.

Every time something provoked him to visualise a picture, he consciously threw the image out of his mind. In a way, he chose to become blind in his mind too. Hull's justification was that this was his way of honouring the truth, by not creating a fiction in his head about what the world looked like. Any picture that would pop into his mind—no matter how detailed a description he could find through words, touch or smell—was, he thought, never going to be real.

For those of us who can see, it is hard to imagine what a world—both in reality and in your mind—would be like devoid of images. Even his wife, who though sighted had stood through Hull's blindness for all these years, couldn't understand what Hull felt like. One day she asked Hull, "What do you think about when you think about our son Joshua?"

"Joshua is that giggling, jumbling bundle of boyhood that I throw over my shoulder. He is those tiny feet kicking me in the chest. He is that beautiful, warm face that I lay my hand on as I sleep," he told Radiolab.

Zoltan Torey was shocked to read about Hull's solution to his blindness. An industrial accident made Torey completely blind at a young age and he was asked—like many other blind people—to focus on his other senses and forget about vision. However, he could not agree with his doctor's or Hull's recommendation. Instead, his way of dealing with the blindness was to consciously imagine as many things as possible.

"The reconstructions are so vivid for me that I actually see it," he told Radiolab. Even though Torey has never seen her, this is how he describes his wife's eyes: "They are brown, slightly flecked with yellow spots in it. They are also large and expressive."

It's not that his wife told him that her eyes were expressive. Instead, he says, through years of experience of living with her and hearing about her from others, he has constructed this highly detailed image in his mind. Torey argues that humans are visual creatures that cannot emotionally react properly to things without visualising them.

There may be scientific support to this reasoning. For instance, the retina in our eyes is actually part of the central nervous system that contains the brain and the spinal cord. It even has five types of specialized neurons that take in different kinds of information and feed it directly to the brain. Among all our senses, vision has the largest area of the brain dedicated to understanding what we see. There is an area, called the occipital lobe, which is solely dedicated to processing the heaps of information flowing in from our eyes. Once that information goes through the brain, it travels to other regions, where what we see is assigned meaning.

Between Hull and Torey, we have two different theories about the importance of sight and how else one might think about seeing, in the absence of the function of sight. Hull accuses Torey of "visual totalitarianism", while Torey considers that, in doing what he has, Hull has let his visual self go.

Most of us are fortunate enough to have never experienced what Hull and Torey have gone through, but this is also why it is hard for us to imagine what it must be like to live their lives. There is another way to empathise with such lives though, and it is through the rare stories of those who have been given sight after being born blind.

Torey and Hull's view of the world. Image credit: Ian Morris.

2.2 SHAPING LIGHT

Nearly 350 years ago, the Irish politician and scientist William Molyneux posed a daunting question to the British philosopher John Locke in a letter[2]

> Suppose a man born blind, now adult, who has learned how to distinguish by touch between a cube and a sphere of the same metal and about the same size, so that he can tell when he handles them which is the cube and which the sphere. Now suppose the cube and sphere to be placed on a table, and the blind man be made to see. Can he by his sight, before touching them, tell which is the globe, which the cube?

Since being published in Locke's remarkable 1689 treatise "An essay concerning human understanding," the problem has divided philosophers. The first camp argues that shapes have innate qualities that are shared by our senses. The second argues that to understand roundedness, a person has to see round objects.

Those who can answer the question are people born blind but whose vision has been restored. Until recently, there were no more than a handful of such cases. Since 2003, however, things have been different. The not-for-profit Project Prakash (Hindi for "light") has performed more than 200 surgeries to restore the sight of children between the ages of 8 and 17. Pawan Sinha, professor of vision and computational neuroscience, has supervised these surgeries. Each patient's experience of regaining sight was recorded in detail for a short period after the surgery and some were followed for longer.

"The moments immediately following bandage removal are not quite as 'magical' as Hollywood movies would have us believe," Sinha told the New Yorker.[3] One simplistic way to describe it is to consider what happens to a person with a normal vision when you suddenly expose them to daylight by pulling back curtains. What a blind person sees is a confusing mess of brightness, blurred shapes, and mixed up colours.

In one of the tests, within the first 48 hours of their sight being restored, these children were shown 20 small objects, which they were not allowed to touch. They were also given 20 identical blocks to touch under the table, but kept them hidden from sight. When asked to match two objects by touch alone, they got it right almost every time. However, when they were asked to match an object they were touching with one they could see, they did no better than chance.

So the answer to Molyneux's problem would appear to be "no". Visually speaking, a cube and a sphere are not what a blind person thinks they are when he or she couldn't see them. Although such people quickly adjust to the normal world recognising shapes among other things, they rarely have the kind of 20/20 vision that a person with the same biological set of eyes would have. This is because, even though the brain is plastic to a large extent, it is not able to adapt perfectly to such a vast new source of sensory information.

Despite decades of research, we are only at the beginning of learning how vision affects our brain. As with most scientific endeavours, it is the absence of something that teaches us what is important. So, instead of studying people with species typical vision, it is the world of the partially blind—who can see some things but not others—that can tell us more about what sight involves.

2.3   INCOMPLETE PICTURES

Imagine a life where the person in the mirror is not you but a stranger, where something as simple as scratching your head and watching that stranger do the same throws you off, where only after many painful seconds of making weird faces do you realize that the mirror is a reflection of your face. Such an experience is routine for extreme cases of prosopagnosia, or face blindness, a condition that was described in the 1950s. Most of these patients show lesions at the bottom of the part of the brain called the occipitotemporal cortex. In fact, the neurons in that area are so crucial to face recognition that they are now called the "fusiform face area".

These "face cells" are developed through experience, a phenomenon which is shown by babies' ability to recognize faces at birth, or soon after. Within six months, babies have become so accomplished at such tasks that they are also able to differentiate between the faces of different species. However, they lose this superpower by the age of nine months, if they are not continuously exposed to faces of that species.

Renowned neuroscientist Oliver Sacks suffered from face blindness himself and wrote about an intriguing case in *The Man Who Mistook His Wife for a Hat*.[4] Dr P had a severe form of agnosia that stopped him from recgonising faces or expressions, but also objects—as the title of Sacks' book describes. Yet, despite the limitations, most face-blind people lead relatively normal lives. They learn to recognize people not by how they look, but by other, contextual information. Here is how Sacks describes his experience:[5]

"I avoid conferences, parties, and large gatherings as much as I can, knowing that they will lead to anxiety and embarrassing situations, since I not only fail to recognize people that I know well but tend also to greet strangers as old friends. I am much better at recognizing my neighbors' dogs (they have characteristic shapes and colors) than my neighbors themselves. But if I should pass either woman on the street without her dog she might as well be a complete stranger."

Then there are those who see the world differently. Nearly one in 20 people suffers from colour blindness. This occurs because of abnormality in the retina, and the result is a world that has far fewer colours. Most colour blind people have trouble recognising colours on the red to green spectrum. For instance, many understand what do at a traffic light only because of the relative position and not the colour of the light.

In 2002, Don McPherson accidentally found a solution to this problem of incomplete pictures. He had developed glasses that could protect a surgeon's eyes from lasers, but his colour blind friend was blown away by how the world looked different to him through those glasses. So dramatic was the difference that McPherson spent the next several years perfecting it into a product.

Since 2012, EnChroma's glasses have given many such people some really emotional moments. Videos posted on YouTube that capture these first moments have garnered millions of views. People can't stop gaping at the world they have never seen before. The feeling, EnChroma warns its users, can be overwhelming. Tears and shrieks are common features.

In one such video,[6] while being amazed Ethan Scott finds a bottle of disinfectant wipes and asks, "Is this purple?" His reaction to being told "yes," is "Oh my god. What the f***? Holy s***."

The glasses work for only a few colour blind (mostly red-green type of blindness) and they are not a cure. In the same way that normal glasses aid the near-sighted or far-sighted, EnChroma glasses trick the color blind people's brain into seeing images it wasn't able to before. This works because the retina has three types of cones, which detect red, green, and blue colours. Among those with red-green type of blindness—called dichromacy—the red and green cones are overlapped. EnChroma's technology filters out some of the wavelengths of light that lie in the overlapped region. This filtering makes the brain treat the reds separately from the greens, and thus allows for the wonderful reactions.[7]

However, there are rare cases for whom the world simply remains black and white. This is caused in people who don't have cone cells, or if all their cone cells are functioning improperly.

Self-described cyborg Neil Harbisson is a case in point. His vision was one such rare case, and he came up with a unique solution to overcome the limitation. Working with a university student in 2003, he built a device that converted colours into sounds. Once he had memorised how the colour spectrum sounded and "seen" the world with this enhancement for a while, he says that he could go around "hearing" the sounds of colours all the time without feeling troubled.

Subsequently, Harbisson had an antenna implanted in to his skull, which had a little camera at one end and an image-to-sound computer that transmitted the sound *via* bone conduction. The connection with his device was so deep that he even started dreaming in colour, where, instead of the device creating the sounds, his unconscious mind was creating sounds from colours.

Such an extreme solution may not be available to most people, but there are alternatives. While working with colour blind people, Dutch scientist H. L. de Vries stumbled upon something remarkable. His male subjects all possessed two normal cones and a mutant one (red or green). Out of curiosity, he tested the daughters of his subjects. To his surprise he found that the daughters had three normal cones and a mutant one, presumably one that they inherited from their father.

As the daughters had four cones, he called the condition tetrachromacy. He presumed the daughters would be able to see more than normal people, but his preliminary tests didn't show any special results, until 2012.

After 25 years of work with tetrachromatic women, Gabriel Jordan, a neuroscientist at Newcastle University, found someone who could make use of her "super vision." However, to make this discovery, Jordan had to build a new test. With the help of a computer program, subjects were shown three flashes of coloured light. To a normal person—that is, trichromatic—all three colours would appear the same. However, for a tetrachromat, one of those three would appear different because of the way the computer created those images. So, instead of purple, the person would be able to distinguish the colour as a mix of red and green.

Jordan tested 25 women. Only one—codenamed cDa29— succeeded to spot the odd one out in every question. "I was jumping up and down," Jordan told *Discover*.[8]

Theoretically, a person with three functional cone cells could see up to a million colours that would result from the mixture of red, blue, and green. A person with four cone cells, should be able to see 100 million colours (or 99 million colours more than a normal person.)

Sadly, however, if they could, it would be hard for them to describe, since our language has not caught up with such capacities. Most people are trichromatic and our language has only found words for the range of shades that people typically see. Also, given how hard it is to describe the colour red to a colour blind person this person with "super vision" will find it hard to describe the difference she sees among coloured objects that others can't see.

Some people who recognise this capacity do make use of it. One such person is American artist Concetta Antico who uses her powers to paint. For her, a pink rose is not merely pink. It has shades of gold, yellow, orange, purple, red, and, of course, pink.

Research into tetrachromacy hasn't received much attention, so we are still not sure how many people might have this power. As color is such a personal experience, many of them might be walking around without knowing that their world is a little bit more spectacular than others.

The story of light particles is a wonderful metaphor for human lives. The stories of blind, partially blind, and those with super vision tell us about how light shapes our lives, regardless of one's functional capacity. Light pervades all aspects of life and always leaves its mark on the universe.

Indeed, from the very beginning of the universe, light particles have shaped the existence of life on earth. Such particles were created moments after the Big Bang, as our universal forces came into being. Since then, these particles have bobbled around the universe for billions of years. Some have died out, and new ones have been born. When the universe comes to an end, many billions of years from now, some of these light particles will remain. In the words of physicist Kevin Pimbblet (see Chapter 2), these light particles would be so far apart that for all practical purposes they won't exist. Yet, in the absolute sense, they will remain until the very bitter end.

REFERENCES

1. R. Krulwich and J. Jad Abumrad, Seeing in the Dark, Radiolab, WNYC Radio, New York, 2012 October 12 [cited 8th March 2016]. Available from: http://www.radiolab.org/story/245482-seeing-dark/.
2. J. Locke, *An Essay Concerning Human Understanding*, 1689, Prometheus Books, New York, 1995.
3. P. House, What people cured of blindness see, The New Yorker, 28th August 2014 [cited 8th March 2016]. Available from: http://www.newyorker.com/tech/elements/people-cured-blindness-see.
4. O. Sacks, *The Man Who Mistook His Wife for a Hat*, Picador, New York, 2011.
5. O. Sacks, *Face-blind*, The New Yorker, 30th August 2011 [cited 8th March 2016]. Available from: http://www.newyorker.com/magazine/2010/08/30/face-blind.
6. Colorblind man sees purple for the first time. [video file]. 2015 June 30 [cited 8th March 2016]. Available from: https://www.youtube.com/watch?v=WCcxwieuDH0.
7. C. Martin, *EnChroma's Accidental Spectacles Find Niche Among the Colorblind*, The New York Times, 15th August 2015 [cited 8th March 2016]. Available from: http://www.nytimes.com/2015/08/16/business/enchromas-accidental-spectacles-find-niche-among-the-colorblind.html.
8. V. Greenwood, *The Humans with Super Vision*, Discover, 18th June 2012. [cited 8th March 2016. Available from: http://discovermagazine.com/2012/jul-aug/06-humans-with-super-human-vision.

CHAPTER 3

How Has Light Evolved?

Kevin Pimbblet

University of Hull, UK

> Now that we've explored some of the psychological impacts of
> light and colour, it is time for some physics and a question
> about darkness, which reveals something about the nature of
> the universe.

3.1 WHY IS THE NIGHT SKY DARK?

At some point, nearly each and every one of us has stood in the
cold of night looking up at the dark night sky. When doing so, we
naturally look for the light sources to guide our sight: the stars,
galaxies, planets, shooting stars and comets. Depending on
where you are in the world, the night sky can look very different
at different times. However, one thing is always the same—most
of the sky is dark, but why? Why is the sky not filled with
light from all the stars in the universe? It is not even filled with
light that is outside of our electromagnetic spectrum. Why does
it not glow at infra-red wavelengths, or with ultraviolet radiation,
the kind that many of the most luminous stars in the universe

A Flash of Light: The Science of Light and Colour
Edited by Mark Lorch and Andy Miah
© The Royal Society of Chemistry 2016
Published by the Royal Society of Chemistry, www.rsc.org

produce? With so much light in the universe, why is there darkness at all?

Scientists and astronomers have been asking this question for hundreds of years, since 16th century Thomas Digges, a man noted in history for having first translated the works of Copernicus into English. In 1576, Digges concludes that the darkness is due to our inability to see the infinitude of the universe, where stars fall into obscurity from our vision,[1] a view later espoused by Edmond Halley in 1720. Alternatively, at the time of Galileo, the 17th century astronomer Johannes Kepler proposed that the universe of stars was finite, and what we see is the expanse beyond it, a kind of nothingness, but even this didn't fully make sense. Even the poet Edgar Allan Poe anticipated some aspects of the possible answer to this question in his 1848 essay 'Eureka: A Prose Poem'. In doing so, he introduced one key element to the question—the idea that light travels through time and that there must be some light rays which have yet to reach us.

Today, we commonly attribute the mature formulation of the question "Why is the night sky dark?" to the German amateur astronomer Heinrich Olbers, who was born in the Holy Roman Empire in 1758. As well as discovering some of the largest asteroids in the solar system, Olbers realised there was a problem with the common conception of the universe at the time, something now referred to as Olbers' Paradox.

To explain, imagine the universe was infinitely old, without a beginning or an end, never changing and full of an infinite amount of stars. If that were true, then the night sky would look very different to what we actually see. There would be a star in every direction, some closer to the Earth than others, but creating a source of light at every point in the sky. Yet, we know that is not the case: most of the sky is dark, so clearly there is something wrong with this picture of the universe.

Today, we have a very different idea of what the universe looks like. Scientists over the past century have assembled one of the most audacious and successful theories in history: the Big Bang theory. This provides us with unambiguous evidence that the universe is not infinitely old but that it had a very definite start point around 14 billion years ago.

In a universe with a definite start date, only a finite number of stars can ever have formed. There might be a huge number of stars, but it is not an infinite amount. Coupled with the fact that the speed of light is also not infinite, this means that not every direction in the sky will contain a star. Some directions simply have nothing. Consequently, when you look out at the cold night sky, most of it looks dark.

3.2 INVISIBLE GLOW

Yet, there is actually far more to space than what we can see with our human eyes. If we were able to see microwaves, the night sky wouldn't appear dark at all. Instead, it would shine with a very faint glow in all directions. This is the cosmic microwave background radiation, the oldest detectable "light" in the universe. In the same way that the night sky's darkness is evidence that the universe had a definite start point, the existence of this background radiation points to it beginning with a big bang.

To understand this further, we need to know a little about the nature of light itself. Many textbooks will describe light as a form of radiation. Some say it is a type of particle called a photon. Others will say light is a wave, a way that energy travels in an oscillating pattern. The reality is that light appears to behave like both a particle and a wave at the same time.

Whichever way we think of light, we know that it is produced when a particle, such as an atom or an electron, becomes excited and then releases a packet of energy. This energy will then travel across space until it it hits and interacts with something else, such as being absorbed by another atom. Whatever the source of the light, it always travels at the same speed (in a vacuum). This speed—299 792 458 metres per second—might seem very fast but it is not instantaneous or infinitely fast. For example, it takes light an average of 8 minutes and 17 second to travel from the Sun to the Earth. The light from the nearest star to the Sun, Alpha Centauri, takes some 4.73 years to reach us. The red from Betelgeuse, the supergiant that forms the star of Orion's shoulder, takes 650 years to reach us and our nearest neighbouring galaxy, Andromeda, is so far away that it takes light a whopping 2.5 million years to make it across interstellar space to Earth.

This means that looking at the night sky is almost like undertaking some intergalactic archaeology. We see objects that are far away from us at an earlier point in the history of the universe than closer objects. For example, the Sun we see when we look up into the sky is actually how it looked from 8 minutes before. Alternatively, Betelgeuse may have exploded into a supernova centuries ago, but the news—in the form of the dazzling light—just hasn't reached us yet, and if Andromedian aliens were spying on us, they would actually be observing our distant ancestor Australopithecus.

As we peer deeper into the Universe, we looked further and further back in time, but eventually we hit a wall, beyond which we can see no further. This wall is better explained as the cosmic microwave background radiation. There is nothing behind it, as it is the furthest thing we can detect and that means the oldest observable thing. Before this radiation was produced, the universe was opaque—light couldn't travel through it—and so we can never directly see what happened beyond this point. The early universe was essentially filled with a plasma of protons, electrons and photons. This plasma was not only super hot—more than 2700 degrees celcius—but also very dense. Occasionally protons and electrons would combine to form an atom of hydrogen, but then it would be struck by an energetic photon which would shatter the newly formed atom.

3.3   REDSHIFTS AND COLOUR

This opaque universe didn't last long. In the blink of a cosmic eye—a mere 380 000 years after the Big Bang—the expanding universe cooled and photons no longer had enough energy to decouple the electrons and protons. Instead, the photons were able to travel freely around the universe and it is these free floating protons that, today, we can see as cosmic microwave background radiation. However, while we can see this radiation left from the clearing mist of the early universe, it is actually quite tricky to work out how far it has travelled. Consequently, instead of using distances and times as units of measurement, astronomers find it much easier to talk in terms of redshifts, a term which denotes a change to a light's wavelength.

Have you ever noticed how the sound of an ambulance siren seems different when it's speeding towards you, compared with when it's travelling in the opposite direction? As it approaches, the pitch of the blaring sirens seems to increase, but when it races away the tone of the siren seems to get deeper. This phenomenon is known as the Doppler effect, named after the person who discovered and characterised it in 1842, Christian Doppler.

Doppler's insight was that this effect applies not just to sound, but to any kind of wave. He thought that, since light can be treated as a wave, it must also be "Doppler shifted" in this way. In the late 1800s, scientists were able to verify this by studying starlight. They observed how, as the Earth spins, the stars surrounding it appear to be moving either towards or away from us. The light from the approaching stars is closer to the blue end of the spectrum, whereas light from stars moving away is closer to the opposite, red end of the spectrum.

This is because the "pitch" or wavelength of the light changes depending on the direction of its origin to a given subject. To explain this, you can imagine standing on a roundabout with ambulances whizzing around it at close to the speed of light. Those coming into view would appear bluer and those disappearing from sight would seem redder, just as the sound of their sirens would change at the same time. This is because the light and sound waves have been Doppler shifted, as their source moves.

Measuring the Doppler shifts of galaxies gives us yet more evidence that the Big Bang theory is correct. Almost every galaxy in the universe appears redder and so seems to be moving away from us. This suggests that the universe is expanding. However, this conclusion creates a bit of a problem for us. After all, if everything in the universe is getting further away from us, it will all appear redder than it would if everything were standing still. So, what colour is the universe in reality? In order to determine that, we need to know a little bit about how colour images are formed and which objects are producing the light.

We are used to seeing amazing images of the universe thanks to huge telescopes like the Hubble, which launched into low orbit in 1990. However, almost all the pictures that such telescopes produce are different from what our eyes would see if they could actually see that far into space. These telescopes use different coloured filters to capture different wavelengths of light. So, a star whose light appears very red will yield a large amount of light through a red filter but very little through a blue filter.

Some filters can also capture light beyond that seen by humans, for example ultra-violet. The stunning image of far away celestrial bodies are constructed by adding together the images that comes through the various filters.

In astronomy, it is common practise to use redshifts in place of distances. At a practical level, it would simply take too long to measure out distances over a cosmologically significant scale. Even using sonar, or reflected light pulses would take many years to reach our nearest stellar neighbours. Since the Universe is expanding though, we can relate the speed at which galaxies are moving away from us with the distance they reside at. Hubble[2] did this in the early 1900's and calibrated his measurements off known distance indicators to provide strong evidence that this was the correct thing to do. Hence we can confidently use redshift to measure the distance away from the observer.

3.4   LOOK TO THE STARS

Burning brightly in the night sky, stars are the most common celestial object that produce the light we receive down the mirrors and filters of telescopes. In our own galaxy, the Milky Way, there are estimated to be some 10^{11} individual stars shining. In the wider Universe, there is likely to be a similar number of galaxies. Hence, observationally, we could potentially see 10^{22} solar masses of stars out there, if we had powerful enough optics to do so.

However, not all stars are the same. For instance, our own Sun is a fairly typical star and has a yellow-white colour. Stars with less mass than our Sun are redder in colour. At the other end of the mass scale, the most massive stars are blue. Even though the physics of nuclear fusion—the way in which stars shine—remains the same for all classes and types of stars, the physical reason behind the difference in colour is down to how hot they are. In turn, their temperature is determined by their mass.

In everyday life, we can see the effect of temperature on colour whenever we see a bonfire, Bunsen burner, or other hot object. If we heat them up enough, then they glow with a distinct light, the hue of which depends on their temperature. As they grow hotter, they turn from red, through to orange, yellow, white and eventually blue. But that is only in the visible portion of the electromagnetic spectrum. Every human is already glowing in the infrared because we have warm bodies—we just cannot see it. Similarly, stars glow across a wider range of the electromagnetic spectrum than can be seen with the human eye.

As stars age, their colours change. For our own star, the Sun, it will eventually reach a red giant stage in its life some 5 billion years from now. The reason why this will happen is due to how the Sun's core is undergoing nuclear fusion. As the hydrogen is used up, the Sun will develop an inert helium core—that is helium that is not undergoing fusion. The core of the Sun will contract at this point and bring some further hydrogen into the right temperature and pressure range to form a shell of fusing material around its helium core. The increase in the reaction rate will cause the Sun to expand and its outer layers will puff outward. In response, the colour of the Sun will change from a yellowish colour to something that is more red, as its more tenuous outer layers cool off. We therefore have a veritable smorgasbord of colour that different stars can take, which evolves over time for individual stars too.

When we observe other galaxies, we are essentially seeing a combination of 10^{11} different colours from 10^{11} different stars. Since stars are evolving, this superposition of colour from all their internal stars is certainly not a constant either. Astronomers who observe galaxies regularly see a vast array of colours, even accounting for their redshifts. Some galaxies are intrinsically very red. Others yellower. Many are whitish-blue. On top of that, they can even have gradients in their colour across their breadth.

The explanation for these colours is reasonably straightforward though. In some galaxies, star formation has ceased. Presumably the internal gas—the fuel for star formation—has either been used up or been stripped away. In these galaxies, the high mass stars—the blue stars—have died off (they live fast, burn bright and die young, some last just a few tens of millions of years!). All that remains in these galaxies are the low mass stars that are red and exceptionally long lived, with life expectancies longer than the current age of the universe. This ensures that the colour of galaxies that have run out of star-forming gas are red. Meanwhile, galaxies that are currently star-forming are blue. Although they are producing stars of all possible masses, the light output from the high mass blue stars dominates the light output for these galaxies. As such, any galaxy with active on-going star-formation will appear blue.

3.5   COLOUR EVOLUTION OF THE UNIVERSE

Knowing that galaxies are just a big collection of light-producing stars of various colours, scientists are able to determine how this light output and colour can evolve over the history of the Universe. They can also predict how it will change in to the future as well. In 2002, an article was published by a large collaboration of scientists looking at how star formation varied over the course of time.[3] In essence, the collaboration determined the average energy output across a large wavelength range of the electromagnetic spectrum. From their survey of galaxies taken from a very large volume of the Universe, this team of scientists created an average spectrum for the Universe. However, this was only a spectrum, rather than what the human eye might be able to see. To change the spectrum into a colour, the scientists de-redshifted by removing the reddening effect of the expansion of the Universe from each and every spectrum in their survey before summing them up together. Only then did they try to assign a (human) colour to this spectrum. They did this by mapping the spectrum on to the international standard for primary colours[4] to see how the "typical" human eye would respond to and see this light. As it turned out, the the average colour of the Universe is an off-white colour or what was eventually decided to be called 'Cosmic Latte'.

Try this at home

Cosmic Latte doesn't generally appear on the colour palette of software packages but it's easy enough to make by mixing red, green and blue in the right ratios. Open a software package like Microsoft Word or Powerpoint, or even a photography package and create some text. Type 'Cosmic Latte' into the title text box. Now, select the colour change option, find the RGB sliders, and change the values to R255, G248, B231. The new colour of your text will be cosmic latte!

3.6 THE MIDDLE-AGED UNIVERSE

The study of the star formation rate of the universe further demonstrated that the bulk of the star formation in the Universe happened 5 billion years ago. This agrees with experiments conducted by a team that I have been personally involved with,[5] which show that the universe is slowly dying. The peak days of star formation are well behind us and, as the universe slowly fades, its colour changes too. What was once blue is now becoming redder, as the more massive blue stars burn out and die in spectacular supernovae. All that remains are the exceptionally long-lived, low mass, red stars. More than this, the sheer amount of light will also diminish with time. The universe was much brighter in the past than it is now. Such fading will continue as the more vigorously light-producing stars die and give way to the meeker red stars that produce a fraction of a blue star's output.

The middle aged universe. Image credit: Ian Morris.

3.7   THE FAR-FLUNG FUTURE

Given that we are well past the peak of the universe, what does the ultimate future hold for light and colour? Well, we know from a variety of experiments—both historic and modern—that the universe is not only expanding, but the expansion rate is also accelerating. This will have some disturbing effects on light. As the expansion gets ever faster, there will come a point at which we cease to be able to observe far away galaxies—they will be receding away from us at too great a speed. This effect will continue inward toward ourselves. Eventually, we will not even be able to observe galaxies in our own local group—our neighbours will vanish from our view.

On a timescale in the region of 1 to 100 trillion year, the star forming gas that currently resides inside galaxies like our own Milky Way will run dry. At that point, no more stars will, or can, be born. A last generation of stars will fuse hydrogen into helium, as their ancestors did—but they will be the last ever to do so. The bluer stars from this generation will die, followed by the yellower ones. Eventually, with enough time elapsed, even the red stars will finish consuming hydrogen in their nuclear furnaces.

It is estimated that approximately 120 trillion years must elapse before this last generation of stars finally stops producing light from nuclear fusion reactions.[6] After this, the only significant objects that remain in the entire universe will be the end points of stellar evolution—objects such as black holes and neutron stars. Yet, even these will not last forever, as black holes can evaporate due to Hawking radiation, a theory that explains the gradual loss of black hole mass due to the discernment of particles through radiation emission. Although this will not happen until a googol (that's a 1 followed by 100 zeros) years from now, their eventual demise will mean that there will be very little left in the universe at that point.[7]

Yet, even at this point in time, light will still remain, as photons will still be zooming around. However, these photons will have several, interesting differences in their properties compared to those of the present day. Given how sparse the universe will be in the far-flung future, it is very unlikely that these photons will ever interact with anything. Furthermore, their wavelengths will become absolutely livaithanic. The high energy, short wavelength (smaller than the size of an atom) gamma rays from the early universe, have already stretched (*via* redshifting) to longer wavelength (about 2 mm) microwaves. This stretching process will continue until the far future, passing through the long wavelength (about 100 m) radio waves until. At this point in our dark future, ancient, low energy photons will have been stretched to wavelengths longer than the today's observable size of the entire universe! What happens to these photons may necessitate knowledge of new physics as it is plausible that the classic theory of radiation could collapse.

REFERENCES

1. D. R. Danielson, *The Book of the Cosmos: Imagining the Universe from Heraclitus to Hawking*, Basic Books, New York, 2001.
2. E. Hubble, A Relation between Distance and Radial Velocity among Extra-Galactic Nebulae, *Proc. Natl. Acad. Sci. U. S. A.*, 1929, **15**(3), 168–173.
3. I. K. Baldry *et al.*, The 2dF Galaxy Redshift Survey: Constraints on Cosmic Star Formation History from the Cosmic Spectrum, *Astrophys. J.*, 2002, **569**(2), 582–594.
4. G. S. Owen, *CIE Chromaticity Diagram* [document on the Internet]. 1999 June 15 [cited 2016 March 1]. Available from: http://www.siggraph.org/education/materials/HyperGraph/ color/colorcie.htm.
5. S. Driver, *Scientist measure slow death of the universe*. [cited 2016 March 1]. International centre for radio astronomy research. Available from: http://www.icrar.org/news/news_ items/media-releases/the-universe-is-dying2.
6. F. C. Adams and G. Laughlin, A dying universe: the long-term fate and evolution of astrophysical objects, *Rev. Mod. Phys.*, 1997, **69**(2), 337–372.
7. D. N. Page, Particle emission rates from a black hole: Massless particles from an uncharged, nonrotating hole, *Phys. Rev. D: Part. Fields*, 1976, **13**, 198.

CHAPTER 4

Why Is the Sunrise so Colourful?

Chris Arridge

Lancaster University, UK

Our star appearing above the horizon every morning, and descending again at dusk has been used to mark the daily cycles of humanity for millennia. And the colour of the sky at the beginning and end of the day can be as varied as the contents of the day itself. When the hues of a setting Sun are particularly spectacular, we may stop and wonder at the light show. But why can a sunset be so varied and colourful, when the Sun itself is largely unchanging?

We can see the importance of the daily cycle of the Sun in archaeological sites all over the world. One of the most famous is Stonehenge in the UK, which is aligned to the sunrise on winter and summer solstices. Stand in the middle of the monument at dawn on those days and the Sun rises directly between an arch and over the Heel stone, which stands about 100 metres from the rest of the moment.[1]

Similar alignments are also found in prehispanic Mesoamerican civilisations, such as El Mirador in Guatemala.[2] Towers at

A Flash of Light: The Science of Light and Colour
Edited by Mark Lorch and Andy Miah
Published by the Royal Society of Chemistry, www.rsc.org

Chankillo in Peru form a "toothed" observatory to track the movement of the seasons from the rising and setting points of the Sun.[3]

Yet, we don't just look at the skies to track the passing of time. Throughout history, people have looked up at the skies and simply been amazed by its beauty—both at daytime and at nighttime. Sun dogs and halos, produced by the refraction of sunlight in hexagonal ice crystals in the atmosphere, grace our daytime skies, while noctilucent clouds, some 80 kilometres above and on the edge of space light up the night sky as they are illuminated by sunlight, long after the Sun has disappeared below the horizon. Maybe most impressive of all are the northern and southern lights, the aurorae, produced during periods of geomagnetic activity as the solar wind smashes into the Earth's magnetic field. These, and many other phenomena of light and colour can be explained with a few relatively straight-forward laws of physics.

Indeed, it's not just our planet either. Imagine yourself on Mars. You're sitting on a rocky outcrop, watching the closest star sink slowly towards the horizon. The pinkish sky blazes with ochre hue. The planet underneath you is rotating, turning you away from the star at 240 metres per second. As the star inches ever closer to the ground, the sky fades and darkens, rendering the world around you invisible, but revealing the universe above your head.

It's actually surprisingly similar to a sunset on Earth, which really should not surprise us, but how to do they work in terms of light and colour? We can explain these phenomena and why they are similar across other planets by examining the physics that governs how light is scattered and extinguished. These principles apply to Earth's atmosphere as much as those of Mars, Pluto, and even planets around other stars.

4.1   WAVES AND PARTICLES

Our starting point is the entire electromagnetic spectrum, from radio waves through to gamma rays. Stars like our Sun emit light over the entire range of this spectrum and we can describe these different forms of light in terms of their wavelength, which measures the distance between two crests of the wave—just like the distance between two wave crests in the sea. Radio waves have long wavelengths, ranging from kilometres to metres, microwaves come next with centimetre/millimetre wavelengths, then there's the infrared, visible and ultraviolet light with wavelengths measuring down to fractions of micrometres (millionths of a metre). X-rays have even shorter, nano sized, wavelengths and, finally, there are gamma rays, where the distance between the crests of the waves is less than the size of an atomic nucleus. All the time we are bathed in this whole electromagnetic spectrum. Yet, humans can sense just a tiny fraction of it, as appears to us on a rainy day, when the Sun peaks out from behind a cloud to reveal the glorious rainbow, with the red (longer wavelengths) to the violet (shorter wavelengths).

Our Sun emits most of its light in the middle of the visible spectrum, actually peaking near the green part of the rainbow. So why do we see a yellow coloured Sun, a blue sky, reds at dawn and dusk, and why do astronauts out in space perceive the Sun to be white?

Despite the yellow hue of the Sun other wavelengths of light are still present. This is obvious in the rainbow, or the UV light that gives us a tan. So, the possibility of blue is always present, despite the fact that the Sun's light looks mostly yellow from Earth.

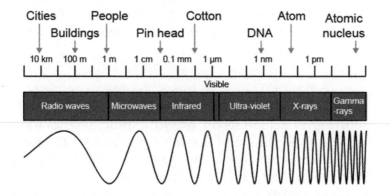

Illustration of the electromagnetic spectrum with typical things at different scales. Image credit: Chris Arridge.

What's more, the colour of the sky, at whatever time of day, is affected by our atmosphere. Light waves from the Sun enter the atmosphere of a planet, then encounter particles and are deflected in different directions. This process is similar to what happens when you're walking through a crowded train station and you have to move to avoid walking into people. This is known as *scattering* and is the cause of many of the light phenomena we know in the sky.

The atmosphere is made of a variety of particles. Molecules of gas are around one 10 000th of a micrometre (μm) across, 100 000 times narrower than a human hair. In Earth's atmosphere the main gases are nitrogen, oxygen, and a number of minor gases including argon and carbon dioxide. Other planets have very different atmospheres. For instance, on Mars, the main gases are carbon dioxide, argon and nitrogen. Around Pluto nitrogen, methane and carbon monoxide dominate.

There are also other particles in atmospheres besides gases. Here on Earth, the stars or the blue of the sky are often obscured by clouds consisting of millions of water droplets. In these clouds, the droplets are around 5 to 50 μm, or about as wide as a human hair, which is a lot smaller than the rain drops we see that fall from the sky. Those are quite a bit bigger, at about 0.1 to 3 millimetres across (100 to 3000 μm). What happens when light scatters off one of these particles depends on the wavelength of the light and the size of the particles, or how big one is compared to the other. The wavelength of visible light is 0.4 to 0.7 μm and light is scattered by molecules of gas in a very different way to cloud droplets and rain.

4.2   SCATTERING AND COLOUR

One of the great achievements of 19th century physics was the unification of electrical and magnetic phenomena to form the theory of electromagnetism. This was led by the work of Hans Christian Ørsted (born 1777), Michael Faraday (born 1791), and James Clerk Maxwell (born 1831). Light waves are electromagnetic in nature, which means they have an electric part and a magnetic part that oscillate together. Light waves are scattered when the electric field in the wave interacts with the electrons in atoms, causing them to oscillate. These oscillating atoms radiate new electromagnetic waves, which combine with the incoming wave and result in a scattered light wave.

When the wavelength of light is much larger than the size of the atom, the atom experiences a smoothly changing electric field and can respond in a slower fashion. This is a little like a boat on the sea gently moving up and down when a wave goes past that has a large distance between the crests of the wave, as illustrated in the following image. The same applies when particles in the atmosphere, such as molecules, are much smaller than the wavelength of light. This resulting scattering process is known as Rayleigh scattering—named after the physicist John William Strutt (born 1842), who would later be known Lord Rayleigh.

When the electric field in the wave makes the electrons in atoms oscillate, the oscillation is stronger for shorter wavelengths of light. In turn, shorter wavelengths of light lead to a stronger scattering. This principle confers the first important property of Rayleigh scattering, which is that short wavelength light is more likely to be scattered than longer wavelengths. Consequently, blue light is more likely to be scattered than red light by a factor of 10. The second important property is that the light is more or less scattered in all directions.

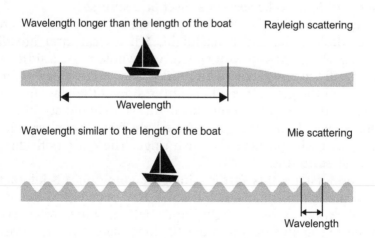

Analogy of how the wavelength of light determines how light is scattered. If the wavelength of water waves is longer than the length of the boat then the boat is rocked in a smooth manner. If the wavelength is comparable or shorter than the length of the boat then the boat will be rocked around very dramatically. Image credit: Chris Arridge.

Rayleigh scattering helps to understand why the sky is blue and sunsets are red. Blue wavelengths of light are scattered in all directions in the atmosphere. So wherever you look, you will see a blue light wave that has been bouncing around from one particle of gas to another. Meanwhile, the red light is scattered less and makes its way to our eyes by a more direct route and so appears to come directly from the Sun.

This even happens during the night—particularly under a full moon—but there is simply less light and so we do not get a very intense blue. In fact, human vision is less sensitive to colour when the light is dim (only the rods in our vision function), so we generally do not see the blue sky at night. The same is true when we look at coloured nebulae through telescopes, the light is sufficiently dim that the cones in our retinas are not triggered.

Sunlight arriving at our eyes from overhead has only travelled around 20 km through our atmosphere. However, when we look near the Sun at sunset and sunrise, the light waves we are looking at have travelled over 500 km through the atmosphere to our eyes. This provides many more opportunities for blue wavelengths of light to be scattered so many times so that it never reaches our field of vision, leaving only the red wavelengths of light to be seen at sunset and sunrise.

You can see this for yourself at home with a glass of water, some whole milk, and a flashlight. Fill a clear straight-sided glass nearly to the top with water. Add whole milk, a little at a time, until the water looks a bit murky. Then shine a flashlight at the side of the glass. If you look at the side of the glass (at right angles to the beam from the flashlight) you should see blue-ish light. Now look into the path of the light from the other side of the class. It should have a red-ish tinge. These are both due to Rayleigh scattering.

To understand this further, imagine going on a 5 kilometre ramble across the moors with friends. Half of you get designated as red wavelengths of light, and half as blue. After every kilometre you all stop and roll a 10-sided dice. If you roll one to nine you change direction and carry on walking, if you roll a ten you keep going on your original route. But the trick is that the people who are red wavelengths of light have weighted dice that roll ten 90% of the time. So, as a red wavelength of light, you are much more likely to keep on route. By the end of the day the chances are half of your 'red' will have completed the route whilst all of the 'blues' will still be out on the moors.

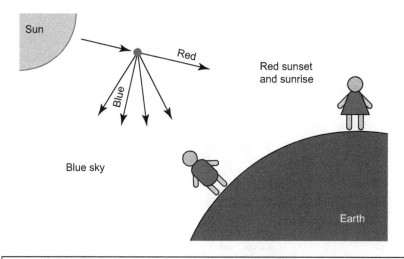

Illustration showing why the sky is blue during the day, but red at sunset. Light from the Sun is scattered by Rayleigh scattering which readily scatters the blue light making the sky away from the Sun appear blue. At some other location on Earth near sunset or sunrise, the blue wavelengths of light will be scattered away leaving only the redder wavelengths. Image credit: Chris Arridge.

As well as dawn and dusk, we also see this effect during total lunar eclipses, when the Earth gets between the Sun and the Moon. The Moon is illuminated by light that has passed through Earth's atmosphere and has been subjected to Rayleigh scattering, removing the blue wavelengths of light. As such, the moon glows red, giving rise to the phrase "blood moon".

4.3 POLARISATION OF THE SKY

A third property of Rayleigh scattering is how it changes the orientation of the light wave. This is known as its *polarisation* and while humans can barely sense it, other creatures are very aware of polarised light. For example, honey bees and migratory songbirds use polarised light to navigate. You're going to meet polarisation again in Chapter 9 on liquid crystal displays and in Chapter 6 on the perception of polarisation.

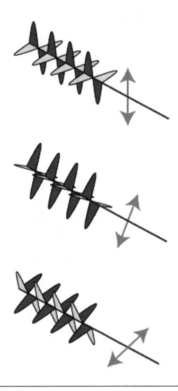

Three light waves with different polarisations. The direction of polarisation is given by the arrow. Image credit: Chris Arridge.

The effect of polarisation may be described as a kind a pendulum, on which we can swing a weight on a bit of string in any direction we like. So, the pendulum could swing forward and back, or it could swing from side-to-side, or somewhere in between. The direction it sways—oscillates or "waves"—is its polarisation.

Now, imagine a group of friends and family together with identical pendulums. Each pendulum is an individual light wave. If you all set your pendulums going in random directions then we would say that they were unpolarised as a collection. However, if they were all going, for example, from side-to-side, then we can say they are polarised in the side-to-side direction.

If we take this analogy back to light, then sunlight comes in as a whole set of light waves that are oscillating in different directions: it is unpolarised. The trick is that Rayleigh scattering selects the polarisation of the light: so the sky becomes polarised. Also, the polarisation at a particular point in the sky depends on where the Sun is in the sky at that time. This means that, at sunrise or sunset, when the Sun is in the east or west, we can tell compass directions based on the polarisation of the sky near the horizon.

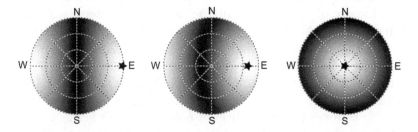

Degree of polarisation of the sky (darker is more strongly polarised) at sunrise, just after sunrise, and at noon. The star symbol shows the location of the Sun. The centre of each circle is straight above in the sky, the horizon is the outer circle with north at the top. At sunrise the sky is strongly polarised in a band running north–south across the sky. At noon the sky is strongly polarised near the horizon. Image credit: Chris Arridge.

The illustration shows the polarisation of the sky at sunrise, just after sunrise, and at noon. It's drawn to show the whole sky with the direction "straight up" in the middle of each circle, and the compass points around the outside. The position of the Sun in the sky is shown by the star symbol. The darker regions show where the sky is the most strongly polarised. At sunrise and sunset the sky is polarised most strongly near north and south.

Migratory songbirds, such as Savannah sparrows, use a number of different senses to navigate, including the Sun, stars, and Earth's magnetic field. Sometimes these senses disagree about the actual direction. This can happen at high latitudes because the Earth's magnetic poles are not in the same place as the geographic poles. When this happens the polarisation of the sky at sunrise and sunset can be used to work out what direction north and south really are.[4] There is also evidence that dung beetles use the same effect at night, when sunlight reflected from the moon is also polarised.[5]

4.4 RED SUNSETS

The second scattering process for light in atmospheres is where the wavelength of light is comparable to the size of the particles. This is the case for droplets of water, particles of dust and ash from volcanoes, and atmospheric pollution, which are collectively known as aerosols. Under these conditions, light is scattered independently of wavelength. In the analogy of a long walk with differently weighted dice, both red and blue groups of hikers would, in this case, have identically weighted dice. As such, they would be equally likely to scatter when they hit an aerosol. This is known as Mie scattering, after German physicist Gustav Mie (born in 1869). Mie scattering can be seen simply by looking at sunlight passing through the steam from a boiling kitchen kettle.

Mie scattering also plays a role in explaining the reddening we see at sunrise and sunset. The red wavelengths of light that are not scattered by Rayleigh scattering make it through the atmosphere to be seen at sunrise and sunset and are further scattered by atmospheric aerosols, making brilliant deep red and orange sunrises and sunsets. This happens at times when the atmosphere is filled with aerosols, such as when there are a lot of droplets in the atmosphere or when there is a lot of volcanic ash in the atmosphere.

"Vesuvius in Eruption" (1817–1820) Joseph Mallord William Turner (1775–1851). Courtesy of the Yale Center for British Art.

Sunsets and sunrises have been a subject of art for thousands of years and record the presence of volcanic aerosols in the atmosphere. William Turner's "Vesuvius in Eruption" reflects the fascination with volcanic eruptions in the 1800s. Although Turner's painting was a depiction of the 79 AD eruption, his subsequent paintings, and those of other artists, provide a record of the reddening effects of atmospheric aerosols. The effects of volcanic eruptions from Tambora, Indonesia (1815), Babuyan Claro, Philippines (1831), and Krakatau, Indonesia (1883), and other eruptions, can all be identified.[6] From our understanding of how light is scattered it has even been possible to calculate the volcanic aerosol content of the atmosphere.

In contrast to Rayleigh scattering, Mie scattering is much more likely to scatter light in the direction it was originally travelling. This is why steam is easier to see when there is a light source behind the kettle, or when dust is illuminated by sunlight streaming through a window or by the light from a projector. The effect, unsurprisingly, is known as looking at "forward scattered" light. It can also be clearly seen on a hazy evening when looking near the Sun (don't look directly at it). Mie scattering equally scatters all wavelengths in the forward direction, so you end up with a glowing halo around the Sun.

Eclipse behind Saturn highlighting the E-ring (the outer-most ring that can be seen in this image). Credit: NASA/JPL/Space Science Institute.

Mie scattering is also an important effect in visualizing the ring systems of planets. Saturn's E-ring is filled with small ice grains (μm in diameter) which can scatter sunlight by Mie scattering. When looking at Saturn's rings from Earth, we can hardly see the E-ring, because Mie scattering doesn't reflect light back towards the Earth. However, Mie scattering does forward scatter light very efficiently, so, if we were to move to the other side of Saturn and look back towards the Sun, then the E-ring would suddenly become visible. This can be seen in images taken by the Cassini spacecraft.

4.5   SUNSETS ON ALIEN WORLDS

Of course, sunsets and sunrises are not uniquely seen from planet Earth. Indeed, scientists have also taken images them on different planets, using robotic spacecraft. Sunset and sunrise images have been taken on Mars since the Viking landers in the mid-to-late 1970s. Quite apart from driving human emotions by seeing the Sun rise and set on another world, they also allow us to probe aerosols in the atmosphere of Mars.

If you lived on Mars one of the first you would notice would be how long twilight lasted for. This is because of Mie scattering of sunlight from all the dust particles in Mars' atmosphere. The physics of this is remarkably similar to the effects of volcanic ash on Earth as painted by Turner almost 200 years ago.

Do aliens love their sunsets too? Image credit: Ian Morris.

Mie scattering is generally more complicated than described above. Oddly-shaped aerosols and differing electrical properties will mean that light is scattered slightly differently from the simple picture of Mie scattering. Models of this scattering can be used to predict what is seen during sunset and sunrise and compared with what is actually seen, such as in the Martian sunset. These tell us what types of aerosols are present—for example ice, water droplets, and dust grains—how much there is of each aerosol and where they are in the atmosphere.[7]

Sunset on Mars on 19 May 2005 taken by the NASA Mars Exploration Rover "Spirit" near Gusev crater on Mars. Credit: *NASA/JPL/Texas A&M/Cornell*.

4.6   USING LIGHT TO FIND LIFE ON OTHER PLANETS

We don't have to be sitting on a planet to watch a sunset. On 14th of July 2015 the New Horizons spacecraft took images of a near-sunset on Pluto, at the outer boundaries of the solar system. Layers of hazes and fog can be clearly seen in Pluto's atmosphere, again, which have scattered sunlight by Mie scattering.[8]

Other wavelengths of light, such as ultraviolet and radio, can be modified as they pass through the atmospheres of planets and, subsequently, used as a tool to probe the atmosphere—and even to detect if one is there. These observations can reveal waves in the atmosphere, the temperature of the atmosphere, the pressure, and whether the body has an ionosphere (the ionised part of an atmosphere).

Layers of hazes seen in a sunset from space as New Horizons flys past Pluto. Credit: NASA/JHUAPL/SwRI.

To understand more about how the emission of light changes over time, scientists have used telescopes to track changes. During August, September and November of 1999, telescopes across Europe and North America focused on a star known as HD209458, located in the constellation of Pegasus. The light they were looking at had left the star 154 years ago, starting its journey in 1845, the year in which Wilhelm Conrad Röntgen, the discoverer of X-rays, was born.

In many ways, HD209458 is a fairly ordinary star, similar to our own sun. However, it was of interest because it was found to periodically get brighter and dimmer, going through a cycle once every 3.5 days. This was the signature of a planet orbiting it. Once every 3.5 days the planet moved in between Earth and HD209458, blocking out some of the light from the star, making it dimmer for a short period of time[9,10] much like a partial eclipse when the moon comes between Earth and the Sun and daylight briefly dims.

The following diagram shows what is seen during an eclipse. If the observer is far enough away from the thing that is moving in front of the star then the observer will be in the penumbra or antumbra. In the penumbra a crescent of the star will be seen, in the antumbra only a disc will be seen. You can visualise this for yourself with two coins. Hold a larger coin at arm's length and then move the smaller coin from your face towards the larger coin until it blocks out the larger coin perfectly. This is a total eclipse and you are basically in the umbra. Moving the smaller coin back towards you and then from side-to-side you will see a crescent of the larger coin and a "ring" of the larger coin as you move the coin from side-to-side. Here you are moving between the penumbra and antumbra.

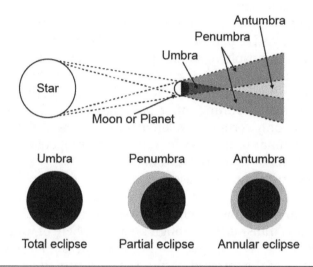

Illustration of what is seen when a moon or planet moves in between a star and an observer. Different sights are seen depending on where the observer is. Image credit: Chris Arridge.

Planets that orbit stars beyond our solar system are known as extra-solar planets, exoplanets for short and they are especially interesting to us in our search for planets similar to Earth. We had known that a planet was orbiting HD209458 because measurements showed that HD209458 wobbled about as the gravity of the planet pulled it around, similar to a child pulling an adult towards chocolate in a supermarket. But this was the first time that we'd seen this partial eclipse—known as a transit. We also see these in our solar system when Mercury or Venus goes in front of the Sun, or when one of the moons of Jupiter goes in front of Jupiter. The next transit of Venus won't be until December 2117, but we will have transits of Mercury in May 2016 and November 2019.

The light from the planet, HD209458b, itself was out-shined by its parent star, making the planet essentially invisible to our telescopes. But the light also passed through the atmosphere of the planet, like light going through a stained-glass window or through a glass of water, and we can use this to study the atmosphere of the planet.

Both our sun and HD209458 have temperatures of around 5500 degrees Celsius and so most of the light is emitted in the middle of the visible part of the spectrum. When this light passes through the atmosphere of the planet, some of it will be absorbed (the starlight will get dimmer overall). Just like when we hold a thin piece of paper in front of a light source, the light will be ever so slightly dimmer after passing through it. But in the atmosphere, only certain wavelengths or colours of light will be absorbed, similar to using coloured sweet-wrappers only letting certain colours of light pass through.

For a given mixture of gases, at a certain pressure and temperature, we can calculate what wavelengths of light will be absorbed, and by how much. The following figure shows how much light is transmitted through an atmosphere of water (H_2O) and oxygen (O_2). We can see that water absorbs a lot of light in certain parts of the infrared part of the spectrum. If we measure the light from a planet transiting in front of its star, see that light is absorbed at the same wavelengths, then we can infer that the atmosphere of the planet contains water. This is the technique of spectroscopy.

Different molecules, like methane and carbon dioxide, absorb light at different wavelengths. So each molecule has a sort of spectroscopic fingerprint that we can use to work out what molecules are in the atmosphere of the planet. We can determine what gases are present and what the pressure and temperature is in the atmosphere. However, the effects of scattering still need to be considered.

Spectroscopy of the atmosphere of the planet around HD209458 showed the presence of carbon monoxide, but not a lot of water.[11] Since then we've found evidence of sodium, water and methane in the atmospheres of exoplanets. Part of this effort is focused on searching for habitable planets, that is those where there is plenty of water in a liquid state, with an oxygen atmosphere. Small amounts of carbon dioxide might indicate life *via* respiration. The analysis of light that has been modified by the atmosphere of a planet is an essential tool in our search for habitable planets, and for searching for life beyond Earth.

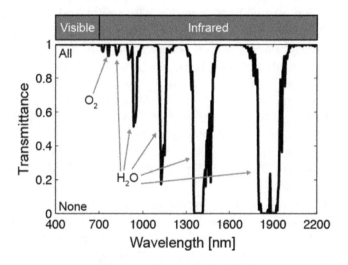

Spectrum for an atmosphere filled with oxygen (O_2) and water (H_2O) showing how much light is transmitted at different wavelengths. When the transmittance is 1 then all of the light will make it through, if it is 0 then all the light will be absorbed. Image credit: Chris Arridge.

4.7 USING LIGHT TO FIND LIFE ON OTHER PLANETS

Sunrise and sunset have a deep meaning for human beings. They mark daily cycles and the passage of seasons. They are also inherently beautiful, whatever planet you are on. The physics of light in the atmosphere of a planet provides an amazing and diverse collection of spectacles. From the volcanic sunsets of Turner, to layers of hazes in Pluto's atmosphere, and Sun Dogs. It also helps humans mark the passage of time, various members of the animal kingdom to navigate around the Earth, and the means for humans to search for potentially habitable planets around distant stars.

REFERENCES

1. J. H. Robinson, Sunrise and Moonrise at Stonehenge, *Nature*, 1970, **225**(5239), 1236S.
2. I. Sprajc, C. Morales-Aguilar and R. D. Hansen, Early Maya Astronomy and Urban Planning at El Mirador, Peten, Guatemala, *Anthropol. Noteb.*, 2009, **15**(3), 79–101.
3. I. Ghezzi and C. L. N. Ruggles, The social and ritual context of horizon astronomical observations at Chankillo, in *Archaeoastronomy and Ethnoastronomy: Building Bridges Between Cultures*, ed. C. L. N. Ruggles, IAU Symposium Proc. Series, 2007, vol. 278, pp. 144–153.
4. R. Muheim, J. B. Phillips and S. Akesson, Polarized light cues underlie compass calibration in migratory songbirds, *Science.*, 2006, **313**(5788), 837–839.
5. M. Dacke, D. E. Nilsson, C. H. Scholtz, M. Byrne and E. J. Warrant, Insect orientation to polarized moonlight, *Nature*, 2003, **424**, 3.
6. C. S. Zerefos, V. T. Gerogiannis, D. Balis, S. C. Zerefos and A. Kazantzidis, Atmospheric effects of volcanic eruptions as seen by famous artists and depicted in their paintings, *Atmos. Chem. Phys.*, 2007, 7, 4027–4042.
7. J. B. Pollack, D. Colburn, R. Kahn, J. Hunter, W. Van Camp, C. E. Carlston and M. R. Wolf, Properties of aerosols in the Martian atmosphere, as inferred from Viking Lander imaging data, *J. Geophys. Res.*, 1977, **82**(28), 4479–4496.
8. A. Witze, Pluto snow forecast poses atmospheric conundrum, *Nature*, 2015, **525**, 13–14.
9. G. W. Henry, G. W. Marcy, R. P. Butler and S. S. Vogt, A Transiting "51 Peg–like" Planet, *Astrophys. J.*, 2000, **529**, L41–L44.
10. D. Charbonneau, T. M. Brown, D. W. Latham and M. Mayor, Detection of Planetary Transits Across a Sun-like Star, *Astrophys. J.*, 2000, **529**, L45–L48.
11. A. G. Inas, Snellen, De Kok, De Mooij, Albrecht *et al.*, The orbital motion, absolute mass and high-altitude winds of exoplanet HD 209458b, *Nature*, 2010, **465**(7301), 1049–1051.

CHAPTER 5

Why Can't You Believe Everything You See?

Wendy Sadler
Science Made Simple, Cardiff, UK

Our mind can play tricks on us making us see things that aren't always there, and physics can be just as tricky. You can't always believe what you see. There's no doubt that what your eyes can see in the world around you can bring immense pleasure—but just occasionally the things your eyes see can fill you with fear and dread. Most horror films and Halloween experiences rely hugely on the ability to unnerve you with apparitions or sudden appearances that seem to defy explanation, but can we always trust what we see? Maybe science can give us an alternative view of these spooky goings-on.

Let me tell you a story about an engineer called Vic Tandy, from Coventry University. Vic's labs were based in an old building adapted from two garages back to back and his work involved making medical equipment for intensive care wards. The lab had

A Flash of Light: The Science of Light and Colour
Edited by Mark Lorch and Andy Miah
© The Royal Society of Chemistry 2016
Published by the Royal Society of Chemistry, www.rsc.org

a reputation for weird happenings, but being an engineer, Vic was not deterred.

One night Vic was working alone and he began to feel increasingly uneasy. The temperature seemed to drop, and he could feel the hairs on the back of his neck standing up. He looked up and, out of the corner of his eye, he saw a vision moving across the lab. As he turned to look directly at it, it disappeared. Unable to find a reasonable scientific explanation, Vic reasoned he may be working too hard and decided to call it a night.

The next day, Vic (a keen fencer) was in the lab early and was doing a few repairs on his fencing foil. He had the long foil clamped horizontally in a vice and he popped out to the storeroom to fetch some oil to treat the blade. When he returned, he noticed the same strange feeling he'd had the night before. His skin was clammy, he felt slightly ill, and, as he turned to his bench, he saw his fencing foil vibrating violently up and down.

At this point, most people would have run for the hills, but not our Vic. As an engineer he began to consider what could explain the blade's dramatic performance. Something seemed to be making the blade resonate, some kind of energy or force. He mounted the blade on a moveable trolley and tried moving it around the lab. He found that, at the edges of the lab, there was very little effect, but as he moved it towards the centre, the movement became much more noticeable. This suggested some kind of vibrational energy in the room and Vic thought the most likely source was one of the machines in the building. Vic asked the caretaker about any machinery running and found that there was an extraction fan that was used to ventilate the paint-spray booth at the end of the lab. When they switched this off, the blade stopped vibrating. So Vic solved the mystery of the energy source, but what about the strange physical feelings and the ghostly person?

We know that different objects have different natural resonant frequencies—that is the frequency they prefer to vibrate. As it happened, the length of the lab was just right to resonate at the same frequency of the extraction fan—which was around 19 Hz. The room was behaving a bit like a long tube, blocked at both ends—like a giant organ pipe—and was creating a standing wave with the maximum amount of energy in the centre of the room. When Vic saw the apparition, he was positioned in the middle of the lab. But, why should the energy from the vibration have made him feel uneasy and how could his eyes see something that wasn't there?

5.1   WOBBLY EYEBALLS

In 1976, some NASA research[1] tested the vibration of human eyeballs and found that most resonate at around 18 Hz—close enough to the frequency of the fan to cause a problem. The vibration in the room would have been too low for our ears to hear, so we call it *infrasound*. However, the inability of our ears to detect it doesn't mean that the vibration can't affect other parts of the body. As well as making the eyeballs vibrate, infrasound can also cause mild hyperventilation that, in turn, causes nausea and sweating. This would explain the unsettling feelings people often report in supposed hauntings. In fact, since this experience Vic Tandy has gone on to investigate a number of 'haunted' places (including Coventry Tourist Information Centre!) and found strong examples of infrasound in many of them.[2]

Try this at home

If you have an old alarm clock radio that has LEDs or some LED Christmas fairly lights hanging around, you can experience first-hand what happens when your eyeballs start moving around.

Stand about 2 metres away from the clock and blow a raspberry as loud as you can. As you do this you will also make your eyeballs vibrate a little bit. An LED is actually flashing on and off very quickly—too quickly for your eyes to see. But if you make your eyeballs vibrate as well then they are taking in slightly different versions of the lights with each flash. This should make the numbers appear to go a bit wobbly and you have effectively turned yourself into a version of a strobe light!

You can sometimes see this happening on a film, if you look at propellers of a plane going around or wheels on a car, and it is sometimes called 'The Wagon Wheel effect'.

There is an online gif where you can try this: http://www. michaelbach.de/ot/mot-wagonWheel/index.html

---

Wobbling eyeballs are not the only way we can see illusions of things that aren't really there. Picture the scene as a thirsty explorer tramps across the desert, hopelessly lost, hungry and dying of thirst and suddenly he sees a lake in the distance. When he reaches it, it disappears. Alternatively, imagine the capsized boat survivor clinging to a raft that has been floating for days and sees land ahead—but she never reaches the mysterious island. Each of these cruel illusions can be due to something called a mirage and they happen due to temperature differences in the air.

The more common type of mirage is the type you see on a hot day that often makes it look like there is water in the distance—this is known as an inferior image as it appears *below* the real horizon.

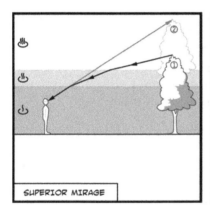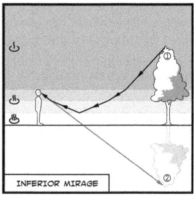

Layers of hot and cold air can cause object to appear above or below where they actually are. Image credit: Liz Bryan Graphics.

Light travels through air at different speeds depending on the temperature of the air. These weird mirage effects happen when you get two (or more) layers of air at different temperatures. Hot air is less dense, so as the light moves into the warmer air it speeds up and moves away from its original path, effectively bending the light. When the light hits your eyes, your brain assumes it has travelled along a straight line so you see a virtual object that seems to be under the horizon. The reason it often looks like water is because, on a very hot day (when the phenomena is more likely), the sky is likely to be clear blue. Our brain interprets the wobbly blue patch beneath the horizon as water.

In a superior mirage, the opposite happens. There is a cold layer of air near the ground with warmer air above it. This is known as a temperature inversion and can often happen in a desert at night (when the temperature drops dramatically), in arctic regions, or at sea. In this type of mirage the illusion appears to hover above the actual horizon and can lead to people seeing ships or icebergs that appear to be floating in mid-air. This is sometimes known as seeing 'castles in the air'.

Most people will have experienced a form of superior mirage without even realising it. When you see the Sun set over the horizon, the reality is that the Sun is already below the horizon. What you are seeing is a kind of mirage of the Sun, because the light is refracted through the different densities of the Earth's atmosphere into our eyes.

The significance of mirage effects cannot be underestimated either. It's thought that a number of 'lights in the sky' UFO sightings could be explained by the fact that lights from cars are refracted above the horizon, so that they appear to be coming from the sky. This could also explain the apparently erratic and unlikely aerial manoeuvres made by the UFOs.[3]

5.2 MERMAIDS, THE FLYING DUTCHMAN AND THE LOCH NESS MONSTER

There is a very specific type of mirage known as the Fata Morgana that is an image changing over time, giving the illusion of movement. In this case, the refraction of the light can appear to vertically stretch small distant objects. The name comes from the Italian name for Morgan le Fay who was the sorceress of King Arthur and Queen Guinevere. These illusions tend to happen in very specific atmospheric conditions and some have suggested they could explain sightings of Mermen and Mermaids at sea as well as the Flying Dutchman ghost ship and the Loch Ness Monster.

If you look at a log floating in the water and the conditions are right, the refraction can stretch parts of the log making it appear as though it is above the position of the actual log. As a result of air movement, this can also create the appearance of the object swimming. Analyses of 'Nessie' sightings backs up this idea because over 77% of monster sightings are between May and

August, when the water will warm up more slowly than the air, giving the required inversion for the effect. Most sightings are also from people down near the water level, which would give ideal conditions for seeing the refraction effect.

The reported sightings of mermaids and merman began to disappear when the decks of ships got higher. Most scientists think that walruses or whales could have been the source of the illusion which, when stretched out vertically and stacked head to toe, could give the impression of shapes that look as though they are half-man and half-fish. The effect would only work when viewed from a position near to the water level. As soon as sailors were viewing from a higher position on deck the stories of mermaids died out. Myths about such occurrences used also to connect the sighting of mermaids with storms coming, or as bad luck for a ship. The connection between these mirages and impending bad weather is also logical as the weather conditions suitable for the mirage often occur just before a storm or unsettled weather.

5.3   INVISIBILITY FOR BEGINNERS

So far, it seems that trusting our eyes may be a risky business, but is it possible to trick them into seeing nothing at all? From superheroes to science fiction, we have long been striving to understand if we could make ourselves invisible and it seems that scientists are getting closer to making this a reality. Most of the successes so far have related to the use of meta-materials which use a particular structure to refract light in such a way that you can see behind an object. In effect, this process involves engineering the mirage idea in a controlled way to give the illusion of invisibility. The meta-material consists of a lattice where the gaps are smaller than the wavelength of light we want to bend.[4] In one case, silk has been used and, as this is not rejected by the human body, there is the hope that it can be used by surgeons to wrap around organs and see behind them. At the moment this is quite experimental and mostly on a very small scale, so how about if you want something a bit larger?

The other way of faking invisibility is to replicate the image of whatever is behind you, a bit like becoming a human chameleon. This technique is known as 'active camouflage'. For example, if you want to develop a Harry Potter style invisibility cloak, then you need a cloak made of a retro reflective surface. To understand more about what this entails, it is helpful to first understand what takes place on normal reflective surfaces. So, a normal surface bounces light at the same angle that it hits, which means that, unless you are standing straight on, you won't see as much of the incident light coming back in your direction. To create the illusion of invisibility you need a different sort of reflection. Retro reflective surfaces are made of tiny balls, which mean that any light projected onto them gets bounced around and reflected back in exactly the same direction that it came.

Cinema screens, traffic signs and cycling jackets use similar material to make them super bright. Finally, the person wearing the cloak needs to have a camera, which is looking at the background. This is linked to a projector situated near the observer, which projects the same image onto the cloak of whatever is behind the person or object.

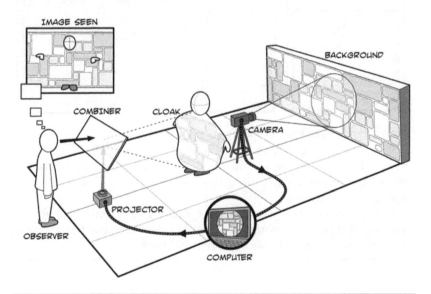

Invisibility cloak. Image credit: Liz Bryan Graphics.

Now, this is not quite a perfected system. After all, the system only allows you to be invisible to a person who is looking at you from a fixed spot, with lots of equipment already set-up! But, there are other applications that are more likely to become a reality, including systems that allow pilots to make the floor of a plane invisible to help them when landing the plane. Such devices may even turn up in our cars to reduce the number of blind spots that exist presently.

DIY Invisibility

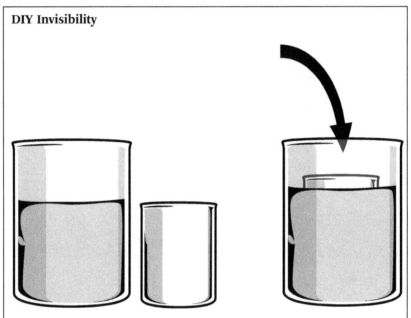

Image credit: Liz Bryan Graphics.

Take a Pyrex jug and an ordinary drinking glass. Fill the jug with vegetable oil or glycerine and then slowly submerge the drinking glass. At the point when the glass is filled with the oil it should appear to disappear!

This is because the glass and the oil have a very similar refractive index (which is the amount the light bends when it enters the object) and so it seems to become invisible in the oil.

REFERENCES

1. M. K. Ohlbaum, *Mechanical Resonant Frequency of the Human Eye 'In Vivo'*. Doctoral dissertain. No. AMRL-TR-75-113. AIR FORCE AEROSPACE MEDICAL RESEARCH LAB WRIGHT-PATTERSON AFB OH, 1976.

2. V. Tandy and T. Lawrence, The ghost in the machine, *J. Soc. Psych. Res.*, 1998, **62**(851), 360–364.
3. J. Naylor, *Out of the Blue: A 24-hour Skywatcher's Guide*, Cambridge University Press, 2002.
4. A. Aliev, Y. Garstein and R. Baughman, Mirage effect from thermally modulated transparent carbon nanotube sheets, *Nanotechnology*, 2011, **22**(43), 435740.

CHAPTER 6

Can We See Beyond the Rainbow?

Juliette E. McGregor

University of Rochester, USA

In this chapter we explore whether it is possible to see colour beyond the wavelength range we usually consider as 'visible'. We discover that many animals can see ultraviolet light and consider how this colour appears to them. We learn that we all have X-ray 'vision' but that it isn't a terribly useful superpower and discover an even more surprising visual ability that most people are totally unaware they have—the ability to perceive the polarization state of light with the naked eye.

The rainbow was first characterised in its familiar 7 colour form by the famous scientist Isaac Newton, not following any careful measurement, but because of his interest in Ancient Greece. The Greeks thought seven was a particularly special number: there are seven days of the week, seven notes on the musical scale, and seven planets in the solar system—or so they thought at the time. Newton wanted to make the spectrum fall into a similar pattern, so he added orange and indigo to his original divisions of red, yellow, green, blue and violet.

A Flash of Light: The Science of Light and Colour
Edited by Mark Lorch and Andy Miah
© The Royal Society of Chemistry 2016
Published by the Royal Society of Chemistry, www.rsc.org

In reality the rainbow is composed of a continuous spectrum of wavelengths and the number of colour hues that the human eye can discriminate is vast. However, just like Newton's enthusiasm for lucky number seven, perhaps our very familiarity with the colours of the rainbow prevents us questioning what lies beyond the wavelength range described as 'visible'. In this chapter, we will discover that there is more to perceiving light than immediately meets the eye and introduce you to some little known visual superpowers of your own.

As Chris Arridge describes in an earlier chapter, light is an oscillating electromagnetic wave travelling through space so it has an amplitude, that we perceive as 'brightness', a wavelength, which we perceive as 'colour' and a third physical property, the plane that the wave is oscillating in—the polarization. Our ability to perceive these properties depends on how these physical aspects of light interact with the visual system. Consequently, it might not be too surprising that parts of the spectrum that we consider 'invisible' are actually accessible to most of the rest of the animal kingdom and polarization cues in the sky are routinely used for navigation. What might come as more of a surprise is that humans can also see beyond the 'visible' range and almost all of us can actually sense the polarization state of light with the naked eye.

Light rays themselves are not coloured, but rather our experience of 'colour' is a construct of our nervous system allowing us to interpret and act on wavelength information from the external world. Colour vision is a vast and complex topic (see also Chapters 1 and 2), but at its base is the conversion of light waves of a particular wavelength into electrical signals, distilling wavelength information into a form that the brain can process. The human eye has a layer of light sensitive cells lining the back of the eye called *photoreceptors*, which perform the first stage of this conversion. Humans have three types of photoreceptors specialized for colour vision (often referred to as 'cones' because of their shape), each of which contains a molecule that absorbs a slightly different range of wavelengths. The sensitivity range of the three cones is shown in the following figure. The short wavelength sensitive photoreceptor has peak absorption at 420 nm—this is often referred to as the 'blue cone'. Additionally we have a 'green' medium wavelength sensitive cone, with a peak absorption at 534 nm and a 'red' long wavelength sensitive cone peaking at 563 nm. With these three photoreceptors we can sense wavelengths from 400 to 700 nm—described as the 'visible range' for humans.

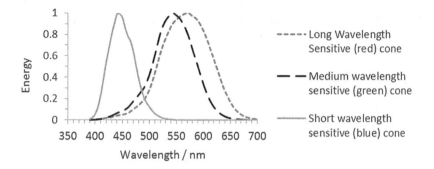

Spectral sensitivities of the three photoreceptors involved in human colour vision. Data source http://www.cvrl.org/cones.htm.

By comparing the stimulation of each of these photoreceptors, the brain can compute the colour. For example, 420 nm, short wavelength light stimulates the blue cone a lot, and the green and red cones only a little; this is the characteristic signature of the sensation 'blue'. Light with a wavelength of 570 nm stimulates both the red and green cones, but not the blue and that is the characteristic signature of what we perceive as yellow. Starting with these comparisons of photoreceptor outputs the human brain can ultimately perceive all the wavelengths of the visible range as the colours of the rainbow.

However, because of our biology, it is also possible to trick the visual system into seeing any colour using just 3 different lights with wavelengths within the sensitivity ranges of each of the photoreceptors. By using a red and green light of the appropriate intensity to stimulate the red and green photoreceptors (but not the blue photoreceptor) in the same stimulation pattern associated with 570 nm light, we perceive exactly the same yellow sensation as before, despite the fact that this wavelength is not actually present. This is the trick that display screen manufacturers use in phones, televisions and computer screens—it is much easier and cheaper to produce just three colours in every pixel and mix them, varying their individual brightnesses, than attempt to produce pixels capable of producing the full spectrum of different wavelengths. So, when you look at an image of a rainbow on a screen, you perceive the whole spectrum of colours, despite there only being three discrete wavelengths present in the image itself. Visual perception is as much a product of our biology as of the light entering the eye. It is therefore natural to suppose that other animals may see the world quite differently to you and I.

6.1 'INVISIBLE' COLOURS?

Humans are unusual in the animal kingdom in that our visible range stops at 400 nm and we don't perceive the shorter, 'ultraviolet' wavelengths. This is because UV from the environment is blocked out by the human lens.[1] Recent research has shown that this is more the exception than the rule in the animal kingdom.[2] Accessing a wider range of wavelengths increases the amount of light and information available and many species take advantage of this. From hedgehogs and okapis, to humble cats and dogs, many mammals have lenses which transmit UV. Reindeer are presumably particularly thankful for their UV vision, as whilst their habitat of snow and ice is highly reflective in the UV, lichen and polar bear fur are not, so these features are high contrast—dark against a brighter background and therefore, more easily spotted.[3] This allows the reindeer to find and enjoy their lunch, without becoming lunch themselves.

So, if there is such an evolutionary advantage, why does the human lens block out UV light? By comparing the eyes of animals, which have UV transmitting and non-transmitting lenses, the explanation appears to be that we have selected for 'high acuity'—the ability to see fine detail—and this has restricted the perceivable wavelength range.[2] Short wavelength UV light scatters more (see Chapter 4) and, as a result, ends up where it does not belong, degrading the contrast of the image we see. The lens also focuses different wavelengths to slightly different depths in the eye, so being sensitive to large a wavelength range could also mean that the UV image would be blurred, even if the red component of the image was in focus on the retina. Whilst our ancestors could probably detect UV, modern humans have evolved to remove this part of the spectrum, allowing them to see fine detail. Without this ability, you would not, for example, be able to read these words!

Whilst most humans are not sensitive to UV, in the early days of cateract surgery, human lenses were replaced by artificial ones which did not block UV, so UV light could reach their photo-receptors, just like the reindeer. The question then arises as to whether these individuals perceived an extended rainbow with an extra totally new wacky colour at wavelengths shorter than blue. Unfortunately not—people with artificial lenses reported perceiving UV as 'a desaturated blue-violet'.[4] UV is just another wavelength capable of stimulating your photoreceptors—there is no change to the human visual system, so the blue part of a rainbow might look a bit wider than it did before, but not that much weirder.

However there are other animals which might have the ability to perceive UV as a colour that we can't even imagine. Birds, fish, reptiles and amphibians have an additional photoreceptor dedicated to detecting these very short wavelengths. With four photoreceptors instead of three, the colour space constructed by their brains, potentially has an extra dimension compared with our own. It is impossible to know how colours look in an animal's 'mind's eye but we may get an insight into what a four dimensional colour space would be like by thinking about our experience relative to animals who have fewer photoreceptor classes than humans.

Consider animals which only have two types of photoreceptors like dogs and cats—they can see colours, but have a colour space more like people who are, for example, red-green colour blind (see Chapter 1) It is possible to make two mixtures of colours that cats, dogs and colour blind people cannot tell apart, but that humans with three normal photoreceptors can easily separate. The situation is even more restricted for deep sea fish—who only have one type of photoreceptor and therefore no colour vision. Whilst their vision is tuned to be extremely sensitive to the beautiful blue bioluminescence produced by ocean creatures, they only see it as 'light' and will never know that it is 'blue'.

6.2 WHAT IS THERE TO SEE?

If we had goggles to block out other wavelengths or if humans just had a single photoreceptor sensitive only in this range and thus see *only* UV light, how would the world appear to us? UV light is scattered strongly by our atmosphere, but enough UV from the sun still reaches the surface of the earth to give us a nasty burn. There is also enough UV light to reflect from flowers and animals to serve as a communication channel. Examples include UV reflecting scales on female cabbage white butterflies that allow the males to identify potential mates and plants that have evolved UV patterns on their flowers to guide insects to their nectar ensuring that they pick up a heavy dusting of pollen on their way.[5]

After we had finished admiring the UV markings on flowers and butterflies, we would notice that glass absorbs UV so windows would be black, neither letting light in nor letting it out. As a result, our cityscape would be dark at night, as would our houses in the daytime. For the same reason, spectacles would look dark like sunglasses. The sky would be bright, but the sunset would be dark. Since UV light scatters so strongly, shadows would look faint, as they are filled with scattered light from the air. Similarly, we wouldn't see very far, as the abundant scattered light would render everything murky.[6]

How about a UV universe? More massive stars are hotter and peak emission is in the UV (see Chapter 2). S-monocerotis, just west of Orion's belt and south of Gemini, has a mass 59 times greater than the sun,[7] with a surface temperature of 35 000 K,[8] compared to 5500 K of our own Sun. To our own visual system these massive, hot bright stars appear to have a blue tinge—with UV sensitivity or a dedicated UV photoreceptor they would appear even brighter.

6.3   REAL X-RAY VISION

The electromagnetic spectrum by no means ends with UV. Below wavelengths of around 10 nm, roughly the size of a protein, we are into X-ray territory. The thought of 'X-ray vision' has captivated human imagination as the archetypal 'supersense' since the first images of skeletal hands showing that you could use them to look inside things (see Chapter 8). In the popular imagination, this idea often takes the form of X-rays somehow shining out of your eyes and, generally, also involves looking through clothes. When it was first discovered, people were so worried that X-ray proof underwear was being developed for respectable ladies concerned about 'X-ray fiends' and 'oglers'.[9] Thankfully, X-rays do not shine out of anyone's eyes, not even those of peeping Toms, but does everyone actually have X-ray *vision* after all?

It was shortly after the discovery of X-rays that the controversy arose—could you actually see them with the naked eye? Rontgen, the discover of X-rays (see Chapter 8) professed that it was not possible, stating that "The retina of the eye is insensitive to our rays; the eye brought close to the discharge apparatus registers nothing.[10]" It might have been tempting to accept that from the authority on the subject, but in the true spirit of scientific enquiry, another researcher quickly presented a conflicting report of a faint blue grey glow seeming to originate within the eye itself when observing an X-ray source in a darkened room.[9] This caused Rontgen to re-examine his own experiences with X-ray perception. While it is true to say that he had seen also seen blue grey glow emanating from a wooden door separating him from an X-ray source in the room next door, as he had only seen it once, Rontgen was unconvinced that it was real. This begs the question as to what he thought it was, but Rontgen was reputedly colour blind, so perhaps he was used to seeing things differently from everyone else, and just took a glowing door in his stride!

It was later revealed that the X-ray source 'in the room next door' had been one of Rontgen's brighter ones and, by using a more powerful source routinely, observations of X-rays could be repeated and extended. Bossalino tied a light tight blindfold around his head and held up copper wires between the blindfold and X-rays source and, despite the blindfold, he could accurately report their number of wires and their orientation.[9] Humans do have the capacity to see X-rays, but they don't interact with our eye's lens[4] and so are not focused onto the retina. This means that if you tried out Bossalino's experiment for yourself and crossed two wires to make a 'T' shape, you would see something more like a shadow than a detailed image. However, the T would also be upside down since, in the visible range, the physical image projected on the retina by our lens is always upside down. The image is flipped over by post-processing in the brain, so that the orientation of our perceptual map of our environment matches physical reality. However, without the action of the lens, the X-ray shadow is actually the right way up to start up. Consequently, when it is subject to the normal brain flip, we would perceive it as upside down, which just makes X-ray vision all the more confusing.

With X-ray vision everything is upside down. Image credit: Heather Holst.

Another consequence of the seamless passage of X-rays through the lens is that X-ray shadows are visible to people who have cataracts, which severely impair their sight in the visible wavelength range. If you can perceive light and dark you can almost certainly perceive X-rays because the visual pigments of the retina directly absorb them. Fortunately a cataract operation is relatively straightforward these days, so there is no need for these individuals to resort to X-ray vision.

Needless to say, when I said "If you tried Bossalino's experiment" I'm definitely not suggesting that you do. Thankfully, for most people, getting your hands on an X-ray source presents the first difficulty, but the second complication is that exposing the sensitive tissues of you eye to high energy X-ray radiation will damage them. For this reason, real 'X-ray vision' is a little known curiosity, which is probably best left on its dusty shelf. If our X-ray vision is of little practical use, do we have any real visual superpowers? There is one candidate which may fit this category—the ability to discern the polarization state of light with the naked eye. This is a sense that most of us have, but few of us notice or know how to use it.

6.4   DISCOVERING YOUR SECRET SUPERPOWER

It is tempting to think of light as simply rays or, perhaps, as an avalanche of tiny balls, and these are appropriate mental images in certain contexts. However, these descriptions fail to characterise a fundamental aspect of what light really is and how it behaves—its wave nature. We can perform experiments to convince ourselves that light is an oscillating electromagnetic wave, but we can also sense an aspect of that oscillation directly. As described in Chapter 4, as a pendulum swings back and forth, all the movement is in one plane. Alternatively you could think of waggling a skipping rope that is lying on the floor. As you move the skipping rope from side to side, to make a snake like wave, all of the movement will be in the horizontal plane and this wave would be described as horizontally 'polarized'. Similarly as light travels through space the oscillations of the electric field are in a particular plane, but unless you have a laser light source, there is no reason for all of the photons to oscillate in the same plane, so light is generally unpolarized. When light interacts with matter, such as during scattering (see Chapter 4), reflection or passage through a polarizing filter (see Chapter 9), not all planes of oscillation are treated equally; some are absorbed, which means that, overall, there is a preferred oscillation axis and the light beam becomes partially or completely polarized.

Sources of polarized light are actually quite common, certainly in our offices and homes. The LCD screens of computers and phones (although not the new AMOLED screens) produce polarized light (see Chapter by 9). The sky has polarization patterns as a result of the sunlight scattering from the atmosphere (as described in Chapter 4) and reflected light is often polarized too. This explains why Polaroid sunglasses are able to reduce the intensity of bright reflections from the surface of water, snow, cars etc. So, if we are surrounded by polarized light you might be surprised to know that we can not only tell that the light is polarized, but also discern the orientation of the oscillating electric field with the naked eye alone. In doing, so we are looking directly into the wave nature of light with our own eyes.

When looking at a source of polarized light it is possible to see a faint yellow bow-tie, flanked by blue/purple regions, which appears at the point of fixation. If you move your gaze, the bow tie

moves too. This ghostly image is referred to as 'Haidinger's brushes' after the Austrian mineralogist who first described it. This
tells the observer the orientation of the polarized light waves they
are looking at, because the yellow part of the image appears
roughly perpendicular to them. So, if you saw a horizontal bowtie,
the polarization would be roughly vertical. People rarely notice
Haidinger's brushes because they fade in a couple of seconds if
the angle of polarization is not altered on your retina. So, con-
versely, if you wish to refresh it, then you can simply tilt your head
from side to side or rotate the piece of polarizing filter you are
looking at, or maybe rotate your screen. With a bit of practice, you
can even see Haidinger's brushes in the sky. Viewing conditions
are best in the patch of blue sky above your head just before
sunset as this is when the skylight is most polarized. You might
notice that the yellow axis always points toward the sun.

6.5   WHAT CAUSES HUMAN POLARIZATION
SENSITIVITY?

In 1850, a meeting of the British Association took place in
Edinburgh with some of the most distinguished scientists of the
age. The great Stokes described his experiments on Haidinger's
brushes, showing that the effect is most pronounced in blue light,
and the influential Brewster suggested that the effect is caused by
cornea and lens. Then, a bashful looking nineteen year old rose to
his feet and disputed this explanation, and to make matters
worse, he has brought with him a contraption made out of seg-
ments of polarizing filter stuck together to show how his alternative explanation would work. A biographer rather uncharitably
describes the young gentleman as a 'beardless stripling,[11]' but
despite his age and inexperience, his explanation was spot on.
Haidinger's brushes would just be the start for this particular
beardless stripling, as he was non-other than James Clerk Maxwell
one of the greatest physicists of all time who would go on to unify
our understanding of electricity and magnetism and predict the
speed of light. Polarization was one of Maxwell's fist loves and he
was often found twisting his head from side to side, in a bid to see
Haidinger's brushes in the sky.[11]

Maxwell had made the link between perceiving polarized light
and the yellow 'macular pigments' found in the human eye.

These are 'carotenoid' molecules which we now know absorb blue light very strongly, only if it is polarized parallel to the long axis of the molecule. If we started with white light, and removed the blue part of the spectrum, what's left over stimulates the red and green photoreceptors, but not the blue. As I discussed earlier in this chapter, this is the characteristic stimulation signature of the colour yellow. This phenomenon explains the colour of Haidinger's brushes, but not their shape and orientation—for that we need an arrangement of these pigment molecules somewhere between the incoming light and the photoreceptors.

At the back of your eye is a very small region of the retina, specially adapted to detect fine detail. To maximize the amount of light directly reaching the photoreceptors in this region, non-photosensitive layers of the retina—which would usually obstruct the path of the light—are swept radially outwards. This means that the cables (axons) necessary to pass the signals from the photoreceptors up to neurons for higher level processing, form a radial pattern overlying the photoreceptor layer. It is this radial structure that provides the physical framework for the spatial arrangement of pigment molecules needed to explain the yellow bowtie shape of the brushes.

The pigment molecules arrange themselves perpendicular to the surface of the cylindrical axons, like the hairs standing up on your arms on a cold day. The resulting structure is polarizing filter with circular symmetry. As a result of the radial arrangement of axon 'arms', the yellowish area where the blue polarized light is aligned with, and absorbed by, the pigment 'hairs' is a bowtie shape roughly perpendicular to the orientation of the polarized light. So, a roughly horizontal yellow bowtie would be perceived in response to vertically polarized light.

6.6 THE BIRDS AND THE BEES

The photoreceptors of invertebrate species like bees, moths and beetles have a different design to ours, which make them intrinsically sensitive to the polarization state of the light they absorb. They don't need radial structures and macular pigments. As a result, these animals are more sensitive to this information and can use the polarization patterns in the sky for navigation, even when the blue sky is obscured by cloud. Celestial polarization patterns are not only produced by the scattering of sunlight in the atmosphere (see Chapter 4), but also by moonlight. Researchers have shown that dung beetles use polarization information in the night sky to ensure they are rolling their dung ball in the intended direction.[12] Just as light becomes polarized as it scatters from the molecules that make up the atmosphere, sunlight is also scattered off water molecules. So, if you go swimming, the 'murky' light underwater is partially horizontally polarized. Maybe Polaroid swimming goggles will be the next big thing. Aquatic animals like octopuses and cuttlefish have also been shown to be very sensitive to polarization but just what they might use this information for, if they use it at all, is still an area of active research. Polarization sensitivity in vertebrates (like us) is less common but recent evidence suggests that polarized light may play a role in setting the internal magnetic compass that birds use for navigation.[13]

Octopuses can sense polarization. Image credit: Heather Holst.

Knowing that people, like dung beetles, bees and cuttlefish, can detect the polarization state of light at will, is an entertaining curiosity (and potential source of smugness), but it might not seem like the most helpful superpower outside of a physics class or dinner party. To achieve genuine superpower status, one would really need to be able to use it to do good—rescuing people in peril, for example. Well, maybe we can. Sensing polarized light relies on there being macular pigment in your eye and it being arranged in a pattern over your retina. Low levels of macular pigment have been found in patients with age related macular degeneration (AMD), currently the leading cause of blindness in the developed world. Usually people are only diagnosed with AMD when they start to lose their vision but if we lose the ability to see Haidinger's brushes at an earlier stage in the disease due to macular pigment loss or disturbance of the retinal layers, then people could be treated sooner and the progression of the disease slowed more effectively. We are very far from knowing if this is the case, but efforts are underway to devise a good test, based on the technology that has previously been used to test polarization sensitivity in animals.[14,15] So maybe one day we will find ourselves at the opticians taking an eye test originally designed for octopuses.

The electromagnetic spectrum and our ability to perceive it doesn't begin and end with the colours of the rainbow. Our retinas are in principle sensitive to ultraviolet light, it's just blocked out by the human lens and X-ray vision is not as far-fetched as people generally think. Recent research even suggests that humans can detect wavelengths longer than the visible range, into the infrared.[16] Next time you see a rainbow spare a thought for the deep sea fish who would only see it as shades of grey, and the blackbird who is possibly admiring an ultraviolet arc under your purple one. Maybe ponder how the wavelength information is collected in your own eye for processing into 'colour' and how we can be tricked into seeing rainbows in pictures that are really just mixtures of three individual wavelengths. Beyond colour, humans can also perceive the polarization state of light, so after you have finished examining the rainbow shift your gaze to a patch of blue sky 90 degrees from the sun and tilt your head from side to side and see if you can spot Haidinger's brushes. If you are prepared to brave the strange looks from onlookers, you will be rewarded with the knowledge that you have 'seen' the wave nature of light first hand.

Try this at home: Navigate like a Viking

Image credit: Ian Morris.

It has been suggested[17] that Vikings also used the polarization patterns of the sky as a navigational aid. Crystals of 'Iceland spar' have been found in excavations of Viking longships and tales of magical 'sunstones' appear in their sagas. Were the Vikings just keen amateur geologists or is there a practical explanation for their apparent fascination with calcite crystals? If you pick up a calcite 'rhomb' (like a cube that has been squashed and sheared to one side) from a your local mineral emporium or museum gift shop and place it over some text, you might notice something peculiar; two images of the text appear. This is an effect called 'double refraction and reflects the fact that there are two permitted vibration directions for light travelling through the crystal because of its asymmetrical chemical structure.

What has this got to do with polarization? If your light source is polarized, something even stranger happens. In an orientation that lines up with one of the permitted vibration directions, all the polarized light takes that route and there is only one image, but if you rotate the crystal so the polarized light is neither quite aligned with one path or the other, then some of the light travels in the second path and a dim second image appears. By rotating the crystal, you can change the brightness of the two images until the second image is at its brightest and the other has disappeared. Try this out by looking at your LCD computer monitor through the crystal as you rotate it.

Now it's time to head out into the great outdoors (or your back garden, or the high seas if you want the full Viking experience), stick a small piece of blue tack on your crystal and hold it up to the blue sky just before sunset. When looking through the crystal at the blue tack there should be two images then one, then two again, and so on, as you rotate the crystal. This effect shows you that the skylight is polarized. As skylight is only partially polarized you will probably see changes in brightness rather than total disappearance. Rotate the crystal so the two images are of equal intensity and holding it in this position mark on an arrow on the crystal pointing along a great arc toward the sun. Now you have calibrated your Viking sunstone. Now, even when you can't see the sun, you can rotate the crystal to the two images of the blue tack are of equal intensity and that will define a great arc along which the concealed sun should lie.

If you take two bearings from different patches of sky you can find the sun's position by triangulation. However, beware that not all of the sky is strongly polarized, so sometimes the spot intensities may not change at all. Understanding where to look is part of the skill of being a Viking navigator.

Try this at home: Make your own rainbow

This is easier than it sounds! All you need is a hose with a sprinkler attachment and a sunny day. Rainbows are caused by light entering raindrops and reflecting off the back surface of the droplet into your eyes. Each time a ray of sunlight crosses the air–water boundary, each of the different wavelengths it contains is bent to a slightly different angle. Lenses convert angles to positions so as the rainbow colours entering your eye are travelling at different angles, they are focused by your lens to different positions on your retina; generating the familiar rainbow image. Next time you are watering plants in your garden or washing your car, stand with your back to the sun so that you can see the reflected light from the water droplets and you should find yourself looking at your very own homemade rainbow.

REFERENCES

1. E. A. Boettner and J. R. Wolter, Transmission of the Ocular Media, *Invest. Ophthalmol. Visual Sci.*, 1962, 1, 776–783.

2. R. H. Douglas and G. Jeffery, The spectral transmission of ocular media suggests ultraviolet sensitivity is widespread among mammals, *Proc. R. Soc. London, Ser. B*, 2014, **281**, 1780.

3. C. Hogg, M. Neveu, K. A. Stokkan, L. Folkow, P. Cottrill, R. Douglas, D. M. Hunt and G. Jeffery, Arctic reindeer extend their visual range into the ultraviolet, *J. Exp. Biol.*, 2011, **214**(12), 2014–2019.

4. L. E. Lipetz, The X-ray and radium phosphenes, *Br. J. Ophthalmol.*, 1955, **39**(10), 577–598.

5. R. E. Silberglied, Communication in the Ultraviolet, *Annu. Rev. Ecol. Syst.*, 1979, **10**, 373–398.

6. J. Walker, *The Flying Circus of Physics: With Answers*, John Wiley & Sons, 2008.

7. Z. Cvetković, I. Vince and S. Ninković, A new orbit of binary 15 Monocerotis, *New Astron.*, 2010, **15**(3), 302–306.

8. A. B. Underhill, L. Divan, M. L. Prévot-Burnichon and V. Doazan, Effective temperatures, angular diameters, distances and linear radii for 160 O and B stars, *Mon. Not. R. Astron. Soc.*, 1979, **189**, 601–605.

9. R. M. Swiderski, *X-ray Vision: A Way of Looking*, Universal-Publishers, 2012.

10. W. C. Röntgen, On a New Kind of Rays, *Science*, 1896, **3**(59), 227–231.

11. C. Campbell and W. Garnett, *The Life of James Clerk Maxwell: With Selections from His Correspondence and Occasional Writings*, Macmillan and Company, 1884.

12. M. Dacke, D. E. Nilsson, C. H. Scholtz, M. Byrne and E. J. Warrant, Insect orientation to polarized moonlight, *Nature*, 2003, **424**(6944), 33.

13. R. Muheim, Sjöberg and A. Pinzon-Rodriguez, Polarized light modulates light-dependent magnetic compab orientation in birds, *Proc. Natl. Acad. Sci.*, 2016, **113**, 1654–1659.

14. S. E. Temple, J. E. McGregor, C. Miles, L. Graham, J. Miller, J. Buck, N. E. Scott-Samuel and N. W. Roberts, Perceiving polarization with the naked eye: Characterization of human polarization sensitivity, *Proc. R. Soc. London, Ser. B*, 2015, **282**(1811), 1–9.

15. G. P. Misson, B. H. Timmerman and P. J. Bryanston-Cross, Human perception of visual stimuli modulated by direction of linear polarization, *Vision Res.*, 2015, **115**, 48–57.

16. G. Palczewska, F. Vinberg, P. Stremplewski, M. P. Bircher, D. Salom, K. Komar, J. Zhang, M. Cascella, M. Wojtkowski, V. J. Kefalov and K. Palczewski, Human infrared Vision is triggered by two-photon chromophore isomerizatio, *Proc. Natl. Acad. Sci.*, 2014, **111**, E5445–E5454.
17. G. Ropars, G. Gorre, A. Le Floch, J. Enoch and V. Lakshminarayanan, A depolarizer as a possible precise sunstone for Viking navigation by polarized skylight, *Proc. R. Soc. London, Ser. A*, 2012, **468**, 671–684.

CHAPTER 7

What Colour Were Dinosaurs?

Charlotte Stephenson
University of Hull, UK

The natural world is a colourful place—from the iridescence of a bird's feather, to the patterns on a butterfly's wing. These colours are formed by chemical and structural features at a microscopic level. Colour can play an important role in mating, fighting and camouflage—meaning that displaying colour can be the difference between life and death for many in the animal kingdom. Colours, such as the vibrant pink of a flamingo, can be altered by diet, and the iridescent sheen of a blackbird's feather can change depending on the angle at which it is looked at. But what about the colour of animals that no longer exist? We imagine scaly green dinosaurs, but what if this wasn't the case? What if there was evidence for a fuzzy, ginger dinosaur? Well... now there is! Palaeontologists have studied beautifully preserved fossils at a microscopic level and found tiny features that allow us, for the very first time, to accurately describe the colours of life that existed millions of year ago. This chapter explains how colours are formed in the world we see around us, as well as how things may have looked long, long ago.

A Flash of Light: The Science of Light and Colour
Edited by Mark Lorch and Andy Miah
© The Royal Society of Chemistry 2016
Published by the Royal Society of Chemistry, www.rsc.org

Light and colour pervade all aspects of our world. There's colour in the darkest of corners and the lightest of places. Understanding how colours are displayed in nature and what function it serves, allows an even greater appreciation of colour's magnificence. Indeed, the importance of colour has been unravelled through years of research, but scientists are not only discovering how colours are produced, but how they have evolved in such a way. Today we are able to identify the colours of creatures that lived millions of years ago, allowing scientists to ask questions (and find answers) normally only attainable by biologists studying modern day plants and animals.

Discovering colour in the fossil record can provide an insight into the function of feathers, skin, scales and fur. Scientists still debate the reasons why colour developed, as well as how and when colour found its place in the natural world. Was colour first used for mating, camouflage, temperature regulation? With so many possibilities, the only way we can fully understand the role of colour in the animal and plant kingdoms is to look back at Earth's ancient life.

7.1   GREY FLAMINGOS

To understand colour, we first need to make sense of the idea of pigments, which are one way in which chemically sourced colour is produced in plants and animals. Pigments are chemical compounds found within specialised cells that selectively absorb light from different wavelengths, resulting in their colourful appearance. There are numerous different sources of pigment, depending on the plant or animal you are looking at, but things like skin, eyes, fur and hair all contain pigments.[1] One particularly prominent pigment in the animal kingdom is *melanin*, which is found in *melanosomes*, and melanosomes within cells. These tiny melanosomes are less than 1 micron in size, far too small for them to be detected, individually, by the human eye.[2]

Not only do melanosomes give colour to their host creature, but they also provide photo-protection; a process where harmful UV rays from the sun are converted into heat—something a body finds easier to regulate. Melanosomes come in two main shapes (although shapes do vary slightly between species). First, there are rounded melanosomes (phaeomelanosomes)—we'll call them the meatballs—and second, the elongate melanosomes (eumelanosomes), the sausages.[2] The meatballs are known to give black and brown colours, whereas the sausages result in reds and yellows.

Another source of colouration derives from *carotenoids*—organic pigments found in plants and algae. Carotenoids have an interesting way of working their way up the food chain. For instance, flamingos appear to be pink, but they are actually born grey. Their pink colouration emerges as a result of their diet, which is mostly pink shrimp. In turn, the shrimp get their colour from their diet, which is red algae. Neither flamingos nor shrimps can synthesise carotenoids, but the red algae can. So, the pale shrimp eat the red algae and the grey flamingo dine on the now coloured shrimp. The carotenoids get dissolved in fats, which then move to a flamingo's feather, resulting in their pink plumage.[3]

As animals can't make carotenoids, if these colours are to be found in their skin, hair etc. they must have eaten plants containing carotenoids. However, not all animals can metabolise them to make these colours. Mammals never evolved the metabolic pathway to process most carotenoids. The ability to signal in yellow and red doesn't seem to have been an important factor of a mammal's life-habits—perhaps this is because of their evolutionary origins, as (most likely) nocturnal creatures, meaning signalling in these colours was probably not very useful.[2]

However biologists have found that some forms of carotenoids can be found in human tissues. An example of this is *β-carotene*, found in orange vegetables, such as the humble carrot.[4] If foods containing β-carotene are consumed in huge quantities (and I mean a heck of a lot!), the body is unable to process this carotenoid and deposits it within the skin, causing an orange colouration. Medically, this is known as *Carotenemia*, and although it may seem rather worrying that someone has gone an odd shade of orange, the condition is harmless and will fade over time as the β-carotene is broken down.

Another pigment deserving of our attention is *chlorophyll*—that produces the green colour we see in plants. Chemical bonds in the pigment absorb light of red and blue wavelengths, leaving the green wavelengths to be reflected from the surface.[5] After reading about carotenoids, you may wonder why animals that eat plants aren't green. Well, this is simply because chlorophyll cannot survive the digestion process.

When most people think of weird and wonderful colours in the animal kingdom, an animal that commonly springs to mind is the chameleon. The ability to change colour, most likely for camouflage, occurs due to several layers of *chromatophores*, pigment containing cells. These cells are able to react to chemical changes in the nervous system and bloodstream, altering their spacing and changing colour according to these signals. For the chameleon, chromatophores vary between four key colours (red, yellow, brown and blue), but can mix these to create a vast range of colours—ideal for the chameleon when wanting to blend into the background.[6]

7.2   BROWN PEACOCKS

Pigments are not the only means of colouration, and structural colours work in completely different ways to pigments. Where pigments are chemicals that work by absorbing light at a specific wavelength, structural colours are generated by a physical process of scattering light off very small biological structures. These structures, found within cells, measure a quarter of the diameter of visible light wavelengths. They refract light in different ways depending on their composition and positioning. These structures can also be stacked on top of each other, resulting in a brighter visible colour being reflected.[7,8]

Structural colours can also result in iridescence—the distance light travels is affected by the angle at which it reflects off an object, so depending on the angle at which you look at an iridescent object, a slightly different colour will be seen (a steeper angle will make things look more blue). All of this means that structural colours are much more varied than pigments— different intensities and pure colours. Reds through to ultra-violets. There's no hue that cannot be achieved through structural colours.[7]

Furthermore, there's no rule that says pigment and structural colour can't work together. One example of this is a peacock's feather. The males display their feathers in order to impress the females and find themselves a mate. The pigments in a peacock's feather are actually brown, but their structure and positioning allows the reflection of light, resulting in iridescent blues, turquoises and greens.[9]

Our understanding of the evolution of colour in the natural world can undoubtedly be enriched by studying fossilised creatures from Earth's past—an idea that was inspired by the Victorians.[10,11] Imagine strolling along the Jurassic coastline in the south of England. It's a warm day with a cool breeze, the only sound is of waves gently lapping up onto the sandy shore. Then, from the corner of your eye, you notice something. It's a fossil! But there's something different about this particular fossil, you discover that it's a type of squid, and there, right in the middle of this fossilised squid is an ink sac. It's black, but what you didn't expect is that the sac still contains ink!

INTERESTING FACT: Those savvy Victorians didn't just display their fossil finds on the mantel piece. Once they realised the squid's ink could be extracted, they used the ink to create beautiful illustrations of the very fossil that it came from.

This squid, millions of years after it was buried, and hundreds of years after the Victorians found it, gave scientists a tantalising sense of the colour in the fossil record.

Ink was extracted from the fossil squid and studied using an electron microscope, through which scientists could see lots and lots of small, round balls... not exactly 'meatballs', but very similar looking melanin granules. Even more exciting was that these granules closely matched the melanin of modern squid ink. Upon making this discovery, scientists realised that the melanosome meatballs that cause colour in the natural world today, may well have been around millions of years ago, and that melanin granules had the ability to be preserved in the fossil record.[11]

7.3   GINGER DINOSAURS

The next question was, if melanin was preserved in fossil squid, what other animals was it preserved in? Next to be investigated were fossil feathers. We know melanosomes are found in modern feathers, so scientists hoped they'd be found in fossil feathers, too.

The research was conducted on feathered dinosaurs. These extinct animals had 'proto-feathers', a simplified version of the feathers we see on birds today. Although these dinosaurs had feathers, they didn't use them for flight. Instead, it is believed that their main role was for warmth and plumage display. This most likely resulted in these non-flying feathered dinosaurs looking rather fuzzy.[12,13]

At just over one metre tall and 100 million years old, *Sinosauropteryx* was carnivorous, flightless and 'ginger'! Analysis of this feathered dinosaur from the Cretaceous period has revealed that *Sinosauropteryx* had a prominent row of feathers from the top of its head, right down its neck, to the middle of its back—rather like a Mohican hairstyle. It also had a stripy tail, like a lemur, but orange and white.[10,12,13] By studying the melanosome structures (sausages and meatballs) found within the fossil, palaeontologists found that the dark stripes on the feathers of *Sinosauropteryx* must have been a chestnut/reddish brown colour, and the light bands, which in fact contained no melanosomes, were white.[14,15]

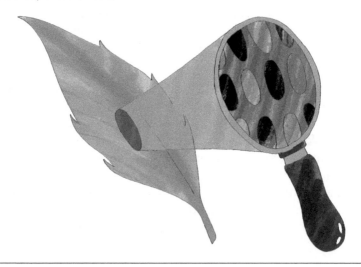

'Sausages' in feathers. Image credit: Heather Holst.

However, it wasn't just melanosomes the scientists found. Fossil feathers, just like the feathers of a peacock, use a mixture of pigments and structural colours. The uniform arrangement of melanosomes within the feathers suggested that structural colour was likely to have played a role in the colouration of *Sinosauropteryx*.[10,14] This arrangement of melanosomes (originally covered with a thin keratin film) would have created an iridescent sheen, which means that, at slightly different angles, the feather would look a slightly different colour.

Microraptor is another example of a feathered dinosaur (like *Sinosauropteryx*), alive 120 million years ago, during the Early Cretaceous. They were one of the smallest types of non-flying dinosaur, weighing about 1 kilogram. The melanosomes found within the feathers of *Microraptor* were found to be rather uniform in their positioning—imagine rows upon rows of sausages, all forming neat lines.[16] In modern birds, a thin layer of keratin covers these melanosome sausages, resulting in a specific hue of iridescence. Unfortunately, because keratin isn't preserved in the fossil record, scientists are unable to define a particular shade of iridescence in *Microraptor* (whether it's more blue, or more black), but due to the arrangement of melanosomes, both pigmented and structural colouration must have been present—just like *Sinosauropteryx*.

Try this at home

Next time you go out, keep your eyes peeled for a feather—a black one would be best. If you're lucky enough to find one, pick it up and move it around in the light. You'll notice, as you move it around, that the colours will change ever so slightly. This occurs due to the interference of light as it travels between the thin film and melanosomes within the feather, as well as the angle at which you look at it.

7.4   STRIPES OR SPOTS

As research continues, scientists are learning more and more about the colours of the past. Since the discovery of melanin in jurassic squid ink, further progress has been made to understand the colours and patterns of a variety of different fossilised creatures. The following examples have yet to be analysed to see if any microscopic structures (such as the sausages and meatballs) are present, but their patterns suggest there's plenty more colour to be found in the fossil record!

While we typically think of colour as a uniform idea, there are plenty of patterns found in the natural world—the spots on a leopard, the stripes on a zebra... but what about patterns in the fossil record?

Perhaps the oldest example of patterns in the fossil record comes from examining the stripes of a 500 million year old trilobite.[17] Trilobites are one of the earliest living arthropods (arthropods are defined as invertebrates with an exoskeleton—other modern examples include scorpions, millipedes and shrimp). This provides an interesting example, as scientists are still unsure as to exactly how developed eyes were by this time. If the eyes of a predator or potential mate couldn't physically see the patterns, then there would seem no evolutionary reason to have them. Perhaps due to light being reflected at different wave lengths, the disruption of UV to another creature's sensors (or primitive eyes) could have been enough to either trigger a warning to predators, attract a mate, or blend into the background?

And this is certainly not the only evidence of patterns in the fossil record. A study conducted in 1989 found 180 different types of animal that had patterns preserved in the fossil record.[17] It has been argued that these patterns may be a result of the fossilisation process (not present when the animal was alive), but if not, prehistoric life could have used these patterns for a number of reasons, including camouflage and warning—the more we study fossils, the more we'll discover.

There are plenty of modern day butterflies that have eye spots, but there's also evidence of this trait from 160 million years ago. These butterfly-type creatures were called Kalligrammatidae, and had a wonderful array of eye-spots, beautifully preserved in the fossil record.[18] Today, butterflies use their spots to either startle or deflect predators such as birds. This was most likely the case with Kalligrammatidae, but in this instance, it was mostly to protect themselves from flying/gliding dinosaurs.

By studying fossils at a microscopic scale, palaeontologists are beginning to unravel the origins of, and reasons behind, colour in the natural world. Identifying the colours of extinct life provides an insight into how animals may have camouflaged themselves to hide from predators, or put on colourful displays to dazzle a potential mate. Up until now, science has had to look at colours from life today, in an attempt to understand colours in the past—making assumptions about how life on Earth must have looked millions of years ago. But now, for the very first time, we can observe colours directly from the past, unlocking the secrets to how colour in the natural world came to be.

> *"Palaeo colour has gone from a world in 50 shades of grey to one painted in Technicolor, and the coming years will certainly see a great number of innovative studies looking at colours of the past."*
> – Jakob Vinther, 2015

REFERENCES

1. G. Britton, *The Biochemistry of Natural Pigments*, Cambridge University Press, Cambridge, 2009.
2. D. Marshal, Colouration in fossils, Paleocast, iTunes, Bristol, UK, 2012 November 15 [cited October 2015]. Available from: http://www.palaeocast.com/episode-7-coulouration-in-fossils/.
3. D. L. Fox and T. S. Hopkins, Comparative metabolic fractionation of carotenoids in three flamingo species, *Comp. Biochem. Physiol.*, 1966, **17**(3), 841–856.
4. E. Kotake-Nara and A. Nagao, Absorption and metabolism of xanthophylls, *Mar. Drugs.*, 2011, **9**(6), 1024–1037.
5. BBC. BBC Bitesize: Leaves and Photosynthesis [document on the Internet]. London; 2014 [cited January 2016]. Available from BBC: http://www.bbc.co.uk/schools/gcsebitesize/science/add_ocr_gateway/green_world/leavesrev3.shtml.
6. F. Stuart, How do chameleons and other creatures change colour? The Conversation [newspaper on the internet] 2013 May 2. [Cited December 2015]. Available from: http://theconversation.com/how-do-chameleons-and-other-creatures-change-colour-13842.
7. N. Dushkina and A. Lakhtakia, in *Engineered Biomimicry*, ed. A. Lakhtakia and R. J. Martin-Palma, Elsevier, Amsterdam, 2013.
8. J. Vinther, D. E. G. Briggs, J. Clarke, G. Mayr and R. O. Prum, Structural coloration in a fossil feather, *Biol. Lett.*, 2010, **6**(1), 128–131.

9. J. Zi, X. Yu, Y. Li, X. Hu, C. Xu, X. Wang, X. Liu and R. Fu, Coloration strategies in peacock feathers, *Proc. Natl. Acad. Sci. U. S. A.*, 2003, **100**(22), 12576–12578.

10. J. Vinther, A guide to the field of palaeo colour, *BioEssays*, 2015, **37**(6), 643–656.

11. D. Marshal, Melanin, Paleocast, iTunes, Bristol, UK, 2015 September 28 [cited October 2015]. Available from: http://www.palaeocast.com/episode-52-melanin.

12. J. O'Donoghue, Feathered dinosaurs show their true colours, *New Sci.*, 2010, **205**(2745), 12.

13. M. Kaplan, Fossil feathers reveal dinosaurs' true colours. *Nature News*, 2010 Jan 10. [cited January 2016]. Available from Nature News: http://www.nature.com/news/2010/100127/full/news.2010.39.html.

14. F. Zhang, S. L. Kearns, P. J. Orr, M. J. Benton, Z. Zhou, D. Johnson, X. Xu and X. Wang, Fossilized melanosomes and the colour of Cretaceous dinosaurs and birds, *Nature.*, 2010, **263**(7284), 1075–1078.

15. C. Sloan, Dinosaur True Colors Revealed for First Time. *Nat. Geo. News*, [document in the internet]. 2010 Jan 27 [cited October 2015]. Available from: http://news.nationalgeographic.com/news/2010/01/100127-dinosaur-feathers-colors-nature/.

16. Q. Li, K.-Q. Gao, Q. Meng, J. A. Clarke, M. D. Shawkey, L. D'Alba, R. Pei, M. Ellison, M. A. Norell and J. Vinther, Reconstruction of Microraptor and the evolution of iridescent plumage, *Science*, 2012, **335**(6073), 1215–1219.

17. D. R. Kobluk and R. H. Mapes, The fossil record, function, and possible origins of shell color patterns in Paleozoic marine invertebrates, *Palaios*, 1989, **4**(1), 63–85.

18. Q. Liu, D. Zheng, Q. Zhang, B. Wang, Y. Fang and H. Zhang, Two new kalligrammatids (Insecta, Neuroptera) from the Middle Jurassic of Dakohugou, Inner Mongolia, China, *Alcheringa*, 2014, **38**(1), 65–69.

CHAPTER 8

How Can We See Inside You?

Benjamin P. Burke

University of Hull, UK

> Ever since the invention of fire we have used light as a tech-
> nology to explore the world around us. Light in all its forms
> allows us to see places that were previously black. This chapter
> guides you through the history and uses of X- and γ-rays and
> how they can show us what cannot be seen easily.

When New Yorker Reed Harris wanted to propose to his girlfriend, he decided to plan a romantic gesture by hiding the engagement ring in his girlfriend's thick milkshake. He only
realised that the plan had backfired when, upon reaching the
bottom of the cup, his girlfriend remained unaware of his proposal.[1] Reed's girlfriend only realised what was going on when
he eventually proposed to her by showing the X-ray that was
subsequently taken of her stomach. It showed that she was indeed in receipt of the ring. If Reed wanted his proposal to be
memorable, then he certainly achieved that.

A Flash of Light: The Science of Light and Colour
Edited by Mark Lorch and Andy Miah
© The Royal Society of Chemistry 2016
Published by the Royal Society of Chemistry, www.rsc.org

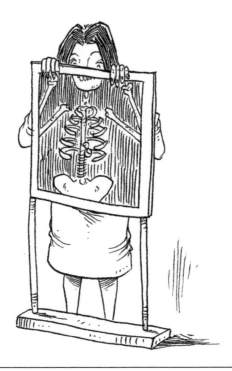

Way to propose! The ring's in there somewhere. Image credit: Ian Morris.

Without the discovery of X-rays, doctors would have had no way to see inside Reed's girlfriend's body and locate the ring—aside from cutting her open and having a look. Until the 20th century, the main way of understanding the inner workings of the human body was by autopsy. Doctors have been using surgery for thousands of years, but poking around inside someone just to work out what was wrong with them isn't exactly ideal and so it was done only in mortal circumstances.

Things have changed a lot since then. Today, by exploiting some of the amazing properties of light, we can now see inside the body, without opening it up, not just examining its basic structure but also visualising organs and diseases in real time. You may have experienced the most basic version of this process, if ever you've had an X-ray image made of your bones or teeth. X-rays are a form of electromagnetic radiation, like light, which passes through most materials. More than 96% of the human body is made up of small atoms (oxygen, carbon, hydrogen and nitrogen) that are not affected by X-rays. However, larger atoms, such as the calcium in our bones, absorb and scatter this energy.[2] By positioning someone in front of a camera and firing X-rays at them from the other side, we can take a photo that shows just the calcium-rich areas, such as bones.

After their discovery in 1895 (see **I have seen my own death**), X-rays quickly became a crucial medical tool, with the first clinical X-ray department opening at Glasgow Royal Infirmary just one year later. As with many technologies, the expansion of the use of X-rays rapidly increased *via* military application, which allowed doctors to visualise, locate and then remove embedded bullets and shrapnel, as well as treat broken bones. The technology was developed further during World War I when the era's scientific superstar, the double Nobel Prize-winner Marie Curie, wielded her political axe in Paris.[3] By October 1916, she had convinced the French government and her powerful friends to support her in converting 20 cars into mobile radiology units, which became known as '*petites Curies*' for the battlefront. She could have stopped there, but not Madame Curie. Not content with merely establishing a new era of battle field medicine she also wanted to be involved in personally deploying it. So she studied the techniques involved in perform X-ray scans, learnt to drive and took her daughter, Irene, aged 17, to the front lines to aid the wounded.

Since being able to visualise bullets in the bodies of soldiers, X-rays have been used to see a wide range of foreign objects in the human body.[4] These run from nails and knives in the human head to "accidents" involving clumsy individuals "falling" on hairspray cans which then became lodged into orifices that have no need for haircare products. Surgical mistakes have also led to some very unusual X-ray scans. For example, in 2014, Karp Ponomaryov from Kazakhstan went to the doctor's suffering a loss of appetite. After X-ray analysis it became clear that something was seriously wrong: the image revealed a 20 cm long pair of surgical scissors left in Karp's abdominal cavity by a stomach surgeon 12 years earlier.[5]

I have seen my own death

In late 1895, German physicist Wilhelm Röntgen was busy exploring the intriguing new technologies of electricity and what could be done with them. He was particularity interested in cathode ray tubes, devices that fire a beam of electrons through a vacuum onto a screen to create images. This is what powered the first televisions. While playing with his cardboard-covered cathode ray tubes, he noticed some strange beams illuminating some nearby screens made from barium platinocyanide. Weeks of dedicated research led to the discovery that these mysterious new beams—which he dubbed "X-rays"—could pass through opaque objects and imprint photographic film behind them.[6]

After studying the X-rays effects on a range of solid objects, he used them to take a picture of his wife Anna's hand that revealed the bones within it.[7] As the first person ever to view a skeletal image of her own hand, complete with the wedding ring on her finger, she was understandably shaken. She reportedly exclaimed: "I have seen my own death." The astonishment and scope of his discovery led Röntgen to receive the first Nobel Prize for Physics five years later. If only Anna had known that the applications of her husband's discovery would improve the lives of so many people.

Seeing one's own death. Image credit: Ian Morris.

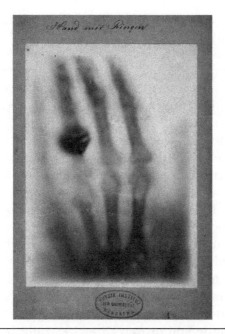

First medical X-ray by Wilhelm Röntgen reportedly of his wife Anna Bertha Ludwig's hand. Image taken by Wilhelm Röntgen.[7]

Try this at home: Make your own X-rays.

Ever wanted to make X-rays at home? Well now you can, sort of. Just take a roll of sticky tape into a pitch black room. Let your eyes adjust to the (lack of) light for a few minutes. Then find the end of the roll (probably should have done that in the light) and rapidly peel the tape off the roll. You should see a flash of blueish light.

Researchers at the University of California at Los Angeles recently showed that if you do the same thing within a vacuum,[8,9] enough X-rays can be produced to make an image of one of the scientists' fingers. The effect is known as triboluminescence and the full reason for why this phenomenon occurs is not known, only speculative explanations have been given so far. If only we had a way to create a vacuum chamber at home, we would be able to make X-ray images in our cupboards at home.

While the breakthrough of X-rays was immediate and near complete, further improvements to medical imaging have been slight and gradual over the 120 years since. The most significant changes have been in improving image resolution but also the development of Computed Tomography (CT), also known as a Computed Axial Tomography, or CAT scans.[10] This technology involves the same fundamental principles as were used back in 1895. However, by rotating the X-ray source and detectors as they operate, while also running detailed mathematical and computational reconstructions, a 3D image is produced. X-rays in their simplest form still allow for easy and cheap diagnosis of injuries such as broken bones, but 3D reconstructions allow for more detailed anatomical visualisation and detection of physiological disease states, such as cancer.

X-rays are far from the only form of radiation that allows us to see inside the body. Other techniques have now been developed that can tell us not just what the inside of the body looks like but how it is behaving. The origins of this advance stretch back to the 1920s, when the English theoretical physicist Paul Dirac[11] proposed that every particle of matter must have an equivalent opposite, a particle of "antimatter".[12] A few years later, American scientist Carl Anderson proved Dirac was right by discovering a new particle which was "something positively charged, and with the same mass as an electron".[13] This became known as an anti-electron or positron.

The problem with this mysterious antimatter is that it no longer exists in any great quantity as it did at the very start of the universe. This might be because antimatter decays at a slightly faster rate compared to matter and so, matter eventually came to dominate. Furthermore, if any particle of antimatter comes into contact with its matter equivalent, the two annihilate each other in a flash of energy,[14] producing two gamma rays which fly-off in opposite directions.

While this makes it difficult to study antimatter, the annihilation itself turns out to be rather useful for medical imaging. This is because each annihilation causes the creation of two gamma ray photons which are emitted in nearly exactly opposite directions. Since X-rays had proved so useful to doctors in the 1950s–1970s, they began to explore whether these high energy cousins of the X-rays, gamma rays, could also be used for imaging the human body. X-ray imaging was great for illuminating static structural features of the body, they were literately photographs. But what if you could create and detect radiation from a substance within the body and study what happens to it in real time? It would be like the transition from photography to the motion picture. Chemists then set to work manufacturing molecules with attached radioactive tags that would decay within the body. In so doing they create positrons which in turn release gamma rays when they encounter an electron. Now remember the gamma ray photons produced by each antimatter/matter annihilation are fired out in opposite directions. When both of these photons are detected you can draw a line, along which the radiolabelled drug must have been. If you then detect a second set (and more) of gamma ray photons you get another line, and where the lines intersect is where your radiolabelled molecules are clustered. This technique is known as Positron Emission Tomography (PET).[11] The trick to making PET a particularly powerful diagnostic tool is to engineer molecules that interact with specific cells in the body. In so doing we are now able to continuously visualise what is happening at a specific location in the body, all by monitoring the gamma radiation released there.

This "molecular imaging" allows us to look beyond the structural features inside the body to specific biological situations. For example, we can study whether a cell has specific features associated with aggressive forms of cancer. The most common method of PET uses a radioactive form of glucose to see which part of the body is most rapidly burning sugar. If one location is consuming significantly more glucose than normal, then this is a good indication of cancer, because tumours need large amounts of sugar to fuel their uncontrolled growth. We can even use radioactive glucose and other positron-producing drugs to probe the functions of the heart and brain.

More detailed medical imaging has had a dramatic impact on medicine and on people's lives. In the case of cancer patients, PET imaging can provide doctors with valuable information on, not just the structure, but the behaviour of a tumour. This information can then be used to inform the treatment. For example, if a patient has undergone chemo or radiotherapy the tumour might not appear to have shrunk any. Such a discovery might suggest an unresponsive tumour and the need for a change in treatment. However, molecular imaging can reveal whether the tumour is consuming less glucose and so, whether it has stopped growing, demonstrating that the therapy is indeed working.

The utility of molecular imaging on patients can be presented with an example. A 63 year old male smoker visits his GP with symptoms consistent with potential lung cancer which was subsequently confirmed by biopsy (physically taking and analysing a small amount of tissue). The standard treatment pathway 20 years ago would likely point to surgery to remove the cancerous region. Instead, the patient is sent for a PET scan with radiolabelled glucose which indicates that the cancer has metastasised (spread) to both the bones and liver. In this case, surgery may have removed the primary tumour it would have no effect on the metastasised regions and likely not lead to an improvement in either quality or length of life. In this instance, based on the PET scan, a better treatment option may be chemotherapy.

In the 120 years since the first discovery of X-rays, the developments and impact on human health have been staggering. Understanding each person's specific genetic situation allows for earlier diagnosis and personalised medicine, helping people live longer and happier lives. How will this change in the next 120 years? Will we have scanners built in to our houses which scan us daily without us knowing? Alternatively, will we abandon medical scanning all together if other blood or urine analysis technologies takes over? What history tells us is that any prediction of 120 years in the future will end up wrong. Whatever method we will be using, it likely hasn't been invented yet.

REFERENCES

1. O. Boyko, Woman swallows diamond engagement ring hidden in milkshake. Daily News [newspaper in the Internet]. 2009 March 3. [cited October 2015]. Available from: http://www.nydailynews.com/life-style/woman-swallows-diamond-engagement-ring-hidden-milkshake-article-1.359627.
2. T. Harris, How X-rays work. How Stuff Works. [Document on the internet] [cited October 2015]. Available from: http://science.howstuffworks.com/x-ray.htm.
3. Marie Curie and the science of radioactivity. The American Institute of Physics, [Document on the internet] [cited October 2015]. Available from: https://www.aip.org/history/curie/brief/06_quotes/quotes_12.html.
4. L. Suen, 25 Strangest Things Found on an X-Ray. IFL Science, [Document on the internet] 2014 Sept 19. [cited October 2015]. Available from: http://www.iflscience.com/health-and-medicine/25-strangest-things-found-x-ray.
5. Amazing World News. Incredible surgeon leaves SCISSORS in man's stomach for 12 years in Kazakhstan [video file].

2014 Dec 10. [cited October 2015]. Available from: https://www.youtube.com/watch?v=KoNRqttCUPg.

6. British Library. Roentgen's discovery of the X-ray. [document on the internet]. [cited 2015 October] Available from: http://www.bl.uk/learning/cult/bodies/xray/roentgen.html.

7. https://commons.wikimedia.org/wiki/File:First_medical_X-ray_by_Wilhelm_R%C3%B6ntgen_of_his_wife_Anna_Bertha_Ludwig%27s_hand_-_18951222.gif H. Markel 'I Have Seen My Death': How the World Discovered the X-Ray. NewsHour Productions. [document on the internet] 2012 Dec 20.]. [cited 2015 October] Available from: http://www.pbs.org/newshour/rundown/i-have-seen-my-death-how-the-world-discovered-the-x-ray.

8. K. Sanderson, Sticky tape generates X-rays. Nature Publishing. 2008 Oct 22. [document on the internet] [cited 2015 October] Available from: http://www.nature.com/news/2008/012345/full/news.2008.1185.html.

9. C. G. Camara, J. V. Escobar, J. R. Hird and S. J. Putterman, *Nature*, 2008, **455**, 1089–1092.

10. US Food and Drug Administration. What is Computed Tomography? [document on the Internet]. Maryland USA. 2015 [cited 2015 October]. Available from FDA http://www.fda.gov/Radiation-EmittingProducts/RadiationEmittingProductsandProcedures/MedicalImaging/MedicalX-Rays/ucm115318.htm.

11. Niels Bohr Institute. PET-scanner. [document on the Internet]. Copenhagen. 2015 [cited 2015 October]. Available from http://www.nbi.ku.dk/english/www/spinoff/spinoff/pet-scanneren/.

12. D. Kwon, Ten things you might not know about antimatter. Symmetry Magazine [document on the Internet] 2015 April 4. [cited 2015 October]. http://www.symmetrymagazine.org/article/april-2015/ten-things-you-might-not-know-about-antimatter.

13. The American Physical Society. This Month in Physics History August 1932: Discovery of the Positron. APS News. [document on the internet] 2004 Aug/Sept. [cited 2015 October] http://www.aps.org/publications/apsnews/200408/history.cfm.

14. Lawrence Berkley National Laboratory. The Particle Adventure. [document on the internet] [cited 2015 October] Available from: http://www.particleadventure.org/eedd.html.

CHAPTER 9

How Do We Make Digital Light?

Mark Lorch
University of Hull, UK

Gases, liquids and solids are not the only possibilities for states of matter, they aren't even the most common states. Plasma, from which stars are made, is much more dominant. The tremendous temperatures inside stars rips electrons from atoms forming gas-like plasma. What's more, there are many other states with exotic names like Fermionic condensates, superfluids and quantum spin liquids, which occur under a wealth of extreme conditions. Others states are commonplace in the world we inhabit and liquid crystals are one of them. This chapter takes a look at this odd state of matter and how it led to light emitting devices dominating our digital age.

The late John Stonehouse has two sets of obituaries written some 12 years apart. The first, from 1974, describes him as a respected government minister in Howard Wilson's Labour government whose untimely death was brought about by drowning, while swimming in the sea off Miami. Newspapers reported that the only trace of him was a pile of clothes left on the beach. There was no sign of his body.

A Flash of Light: The Science of Light and Colour
Edited by Mark Lorch and Andy Miah
© The Royal Society of Chemistry 2016
Published by the Royal Society of Chemistry, www.rsc.org

Stonehouse's second set of obituaries went to press in April 1988 and describe a fallen politician, who refused to resign his parliamentary seat even after he had been arrested on 21 counts including fraud and forgery. They recounted the collapse of his tangled financial empire and his attempt to shirk his responsibilities, by assuming the identity of a dead constituent before faking his own death on that beach in Florida.

None of these obituaries mentioned the classified secret services documents that detail his treacherous dealings with the Czechoslovakian communist state. Nor did they reveal how John Stonehouse had started a ball rolling on a research project that netted £100 million in royalties for the British government. For, despite his many flaws, Stonehouse had a vision in 1967 that the UK needed to develop flat screen, full colour displays. His eureka moment took place at a time when flat screens had only just appeared in such movies as Star Trek. Indeed, Kubrik's 2001: A Space Odyssey didn't manage that sort of foresight until a year later. Nevertheless, Stonehouse instructed scientists at the UK's Royal Radar Establishment to begin construction.

Nearly 40 years later, it is hard to imagine a world without flat screen technology. The 1.7 billion devices shipped worldwide in 2014 allow us to spend almost as much time staring at a screen as we do sleeping.[1] Moreover, the vast majority of those devices to which our eyes are often glued are liquid crystal displays or LCDs. But what is an LCD and how do they work?

9.1   WHAT'S THE MATTER WITH LIQUID CRYSTALS?

Before we get back to the characters, let's start with debunking some misinformation you were told in school. Gases, liquids and solids are not the only possibilities for states of matter, they aren't even the most common states. Plasma, from which stars are made, is much more dominant. The tremendous temperatures inside stars rips electrons from atoms forming gas-like plasma. What's more, there are many other states with exotic names like Fermionic condensates, superfluids and quantum spin liquids, which occur under a wealth of extreme conditions. Others states are commonplace in the world we inhabit and liquid crystals are one of them.

It was quite some time before the first LCD emerged that liquid crystals were observed. In 1888, an Austrian chemist Friedrich Reinitzer was studying a derivative of cholesterol when he noticed something odd, the material seemed to possess two melting points. He started with a solid crystal and heated it.[2] Its form collapsed into a milky viscous puddle, instead of the clear fluid that he was expecting (after all he had observed materials melting many a time, and he'd thought this would behave much like ice turning to clear water). Then something even odder occurred. He applied more heat and observed that the cloudy fluid suddenly turned clear. In this moment, Reinitzer was witnessing a new transition of a material from a solid to a liquid crystal and then to a liquid.

Crystals are solids, the molecules are fixed in place, jammed into an ordered arrangement like a grocer's stack of oranges. If you raise the temperature above the crystal's melting point, then a phase transition occurs and the solid becomes a liquid. The liquid is fluid, it flows and take up the shape of the vessel it is in, the molecules are now free to move around each other, as if the orange stack had been knocked over. Liquid crystals states, as the name suggests, have features of both. They flow like a liquid, but in the free fluid form they also maintain some order like a solid.

The reason for this odd behaviour resides in the shape of liquid crystal molecules. They are usually long and thin. Keep them cool and they will order into crystals, heat them up and the long molecules start to move around but they stay lined up with their neighbours. However, there isn't enough attraction between molecules to hold them in that orientation. So, molecules are free to dance around one another. A great analogy can be seen in packets of pasta. Small bent macaroni is a random jumbled mix, order starts to emerge in penne as the individual pieces line up with one another. Extend the pasta still further, into spaghetti, and it is shipped in perfect parallel packets. But, throw the pasta into hot water and it loses its shape, the constriction of the packet and, hence, the order too.

Long thin molecules are really quite common. DNA, cholesterol (as observed by Reinitzer), detergents and sugars can all form liquid crystalline states. Liquid crystals structures also occur in the membranes that surround our cells where fatty lipid molecules form layers that fold around, forming a sealed bag that defines the edge of a cell. So, how does this fairly common, but largely unseen, state of matter end up as such a fundamental part of our modern devices? A big part of our explanation relies on understanding polarized light and the easiest way to do that is by using waving ropes as an analogy.

Imagine a playground full of children with skipping ropes. They pair up and hold slack ropes between them. Some of them start shaking the ropes up and down, some side to side and others shake it in directions in between. If they all stand together in two lines then their waves might represent a beam of light, but there is no agreement to the direction of the individual waves and their beam is said to be unpolarised. Now, imagine that a teacher instructs them all to shake the ropes up and down. They have now polarized the direction of their waves.

Most sources likes light bulbs, LEDs, or the Sun, generate unpolarised light. But what if you want to polarise it? You simply put a polarising filter in the way that cuts out some of the wave orientations, as one can do with some sunglasses. Sunglasses often have polarising filters as lenses. They work particularly well, because they cut out some of the light, but don't change its colour. Unlike, say, a red-tinted pair of glasses that gives a rosie view of everything. By analogy, let's assume that you have an unruly group of children who ignore instructions to shake their ropes in the same direction. Instead, you feed their ropes through a slatted gate. The parallel slats of the gate only let the up and down waves through. The waves from all the children shaking their ropes in other directions are restricted, and so you've enforced polarisation on the rope waving class.

Unpolarised skipping ropes. Image credit: Heather Holst.

So, what happens if you take a second gate and turn it by 90 degrees. Now, it doesn't matter which way the children shake their ropes, none of the waves can get through. A similar thing happens with polarising filters. Only light waves with a particular orientation are allowed through, and, if you place two filters together, then you don't cut any more light out. However, if you turn one by 90 degrees, together they now turn opaque.

Now, we've got that—and it may help to try my analogy by heading out to a playground and waving some ropes about—lets head back the to the flat screens.

LCDs contain two polarising filters, turned so that no light can get through. Between the polarising filters is sandwiched a thin layer of liquid crystal. When light is shone through the sandwich, half of the light is cut out by the first filter, while the rest would normally be cut out by the second one. However, the liquid crystal in displays does something clever, when sandwiched in between the filters it actually stacks together, quite like the penne or spaghetti, in twisted spiral stacks. As the light passes through the stack, the waves are turned allowing them to slip through the second filter. Going back again to the rope analogy, it is as if another group of children stood in between the two gates and changed the direction of the incoming wave so that it could get through the second filter.

There's one more thing needed to turn this system into a display. You have to be able to switch on and off the liquid crystal's light twisting properties. That way you can flip a pixel of the display from transparent to opaque. One way to do this is to warm up the liquid crystal and melt it transforming it into the liquid. Once that happens the stack of liquid crystal molecules— which is now a true liquid—no longer twists the light. However, warming and cooling tiny areas of a screen isn't really practical. Instead, you need some fine way to disrupt the stack, and that was the crux. It took the, now, infamous Stonehouse to get the ball rolling and crack the problem, but this was only the beginning.

> **Try this at home**
>
> See if you can locate a couple of pairs of polarising filters (they are quite cheap) or sunglasses. You can test if your sunnies have polarising lenses by putting them on and looking at an LCD screen. Then tilt your head from side to side and you'll notice the screen appears to dim and brighten. This is because the light from the screen is already polarised and so your lenses filter some of it out.
>
> Now, on a sunny day, look at a clear patch of blue sky—not at the Sun—and tilt your head back and forth. You'll notice that the intensity of the blue sky appears to change as your glasses filters out the blue polarised light.
>
> One more thing to try. Take two filters and place one over the other, turn one until they turn opaque. Now, take some clear tape and stick it to a transparent sheet of plastic or glass. Cover the glass with stretches of tape stuck down in all directions. Now place your tape-covered plastic between the polarising filters. Suddenly, light once again gets through the second filter. The tape is acting much like the liquid crystal in a display, twisting the polarised light.

9.2   THE SEARCH FOR FLAT SCREENS

In 1963 Harold Wilson aspired to become the first Labour prime minister of the UK in almost 20 years. He delivered a rousing speech at his party conference, that is still written about to this day, calling for a "...Britain that is going to be forged in the white heat of this [technological and scientific] revolution...".[3]

The following year, Wilson's party won the general elections and he took up residence in 10 Downing Street. True to his word, Wilson put in place measures to make the most of the country's scientific advances and formed the Ministry of Technology with Stonehouse at its head.

Some years later, whilst visiting the Royal Radar Establishment (RRE), Stonehouse was shown the large colour cathode ray tubes used for radar displays and was informed that the government was paying the Americans more for the rights to use them than on developing the supersonic airliner, Concorde. This convinced Stonehouse of the need to develop a colour, flat screen panel and he instructed the RRE to get on with it. This was 1967, and so not surprisingly the scientists tasked with the job were horrified at the prospect—after all at the time flat colour screens were literally science fiction. So, instead they set up a working party to hunt for a likely technology. Among the candidates were liquid crystals and ferroelectric ceramics, a kind of smart material discovered in the 1940s.

Several meetings were held in which leading experts in their fields made their case for their technologies to be backed. At the liquid crystal meeting, the expert was asked what appeared to be a straightforward question about why the light reflecting off his sample bottle of liquid crystals cast such a strange pattern on the wall. He began to confidently reply, then stopped and looked puzzled, he flicked through his papers looking for the answer. They tumbled onto the floor where he groveled around desperately trying to find the information he needed to address the question. The chairman of the working group, Cyril Hilsum, started to lose control of the meeting and the audience. At that point, a quiet voice piped up from the back of the room with "I wonder if I can help". To Hilsum's great relief, the young George Gray, a lecturer from the University of Hull, proceeded to provide a beautifully clear answer to the question that had so stumped the session's speaker. At the end of the meeting Hilsum reported back to his superiors that "We must put the man from Hull on the contract". However, there was no contract. The working party hadn't finished its process and still had other technologies to consider.

Finally, in 1969, the report on the future of flat screen displays was ready. Cyril Hilsum sat himself down in a quiet corner of his garden and began to plow through the tomb of a report. Ferroelectic ceramics had triumphed. The group was persuaded that this material was the future of flat screens, while liquid crystals would receive a few crumbs of funding as a backup plan. But Cyril wasn't happy. The report didn't seem to hang together, the logic just didn't flow. So he started doodling. Where he read ferroelectric ceramics he swapped in liquid crystals and vice versa. He read the report again, and this time was satisfied. He returned the doctored report to the group, who, after having spent 2 years on it, were livid, but worn out. The decision to back liquid crystals was approved and ferroelectric ceramics sank into obscurity (if you want to see how the two technologies got on just compare their respective wikipedia pages).[†]

[†]NB. It is well worth watching Cryil Hilsums account of this process during an interview conducted by the British Library.[4]

George Gray received the princely sum of £1400 and started work. Two years later, he had developed a liquid crystal that was stable, easy to manufacture and most importantly worked at room temperature. It was known as 4-cyano-4′-pentylbiphenyl or 5CB for short (the 5 refers to the 'pent' part of the full chemical name).[5]

By 1974 the first calculators containing this compound were on sale and it is still used to this day; if you own a cheap digital watch with a grey and black display you've got some 5CB on your wrist. The colour flat screens envisaged by Stonehouse didn't actually materialise for another 20 years or so, but the government still made over £100 million by licensing the patents for the molecules that Gray had developed.

5CB was a breakthrough for three important reasons. Firstly, it was stable. LCD devices were already on sale a few years before. But, they were incredibly unreliable, the liquid crystal in them reacted with water, which meant that any devices had to be perfectly sealed or even a little moisture from the air would destroy the screen. Ultimately this made them unviable. Second, at ambient temperature, 5CB stacks in just the right way to twist the light. Previous LCD used a slightly different method that involved scattering light, but the screens were slow to respond and had low contrasts, making their performance limited. Lastly 5CB was positively charged at one end, which meant that applying an electric field to the screen pulled on the charged molecule and broke up the stack, turning the pixel dark. Then turn off the power and the stack reforms and the pixel flips backs to white. This meant the pixels could be rapidly turned from black to white.

These elements describe the science behind each and every pixel of your LCD devices. All that needs to be done to make a high resolution LCD is add a backlight, shrink the pixels so you can fit several million of them onto a screen the size of a A4 sheet of paper and, most importantly, be able to control every single one of those pixels independently. The control or address problem (*i.e.* how you send a signal to the correct address on the screen) is a whole other story that has links with silicon chips and solar panel technology. But that story is for another book.

9.3   THE FUTURE IS BRIGHTER

The physical limitations and power demands of cathode ray tubes made them obsolete when compared to the flat and efficient

LCDs. Decades after their initial development, those LCDs are starting to look old too. They are hard and inflexible and they demand rather more power than we would like. The batteries in our portable devices are sucked dry by the light behind our screens. The light has to pass through those polarisers and colour filters before it emerges from the front of the screen. In short, most of the light energy never makes it too our eyes and is, instead, absorbed by the technical requirements of our screen.

The future promises to be much more flexible and efficient. Organic light emitting diodes (OLEDS) are that future. LEDs are already used as the light source at the back of LCD screens. OLEDs improve on this because the pixels emit the coloured light themselves, instead of being filters. So OLED screens need much less power. OLEDs also have one other great advantage, all the electronics could be incorporated into flexible films, which means your future tablet could be rolled up and stuck in your pocket. In fact, examples of such protoypes are already being produced by such companies as Samsung and commercialized in their mobile phones.

Try this at home

There are hundreds of pixels crammed into every inch of a modern LCD screen. Each one is actually made up of a red, green and blue subpixel. Your eye can't resolve each one individually so you see a mixture of the light from all three. So turn them all on and you see white light, and turn on any combination of the subpixels to generate any one of millions of colour.

If you want to take a look at the pixels you just need to put a drop of water on the end of you finger and flick it onto a screen (this works best with older screens as the pixels are bigger). The drops on the screen will act as a magnifying lens revealing the subpixels. BE CAREFUL DON'T POUR WATER ON THE SCREEN.

Take a look at a red part of the screen and you'll see the red pixels are on. Find a purple patch and now you'll see the red and blue pixels, but the green is off. Go to a yellow spot and the red and green are on.

9.4   STONEHOUSE'S LEGACY

After a 68 day trial, John Stonehouse was sentenced to 7 years in prison. He served half of that time, before being released after a heart attack. He married his secretary and mistress, whom he had planned to start a new life with after he had faked his death. She bore him a son and typed up his novels.[6]

In 1988 the Sharp Corporation launched the first commercial LCD flat screen TV.[2,7] Just 1 inch thick and, with its bright 14 inch screen, it heralded a new age of displays. Stonehouse missed seeing his little known legacy by just a few months, as he died earlier that year.

Yet, his story is one of surprising innovation and a tale about how design is often realized by some circuitous routes. The history of science and technology is often like this and it is only by delving into the complex stories of how innovations, patents, and designs are made through the individual and collective actions of complex relationships, that one realizes just how much our present world of technology is made by chance. Stonehouse's story is why predicting the future is hard, but also why so much revolves around artefacts that are focused on the interest to reproduce light and colour.

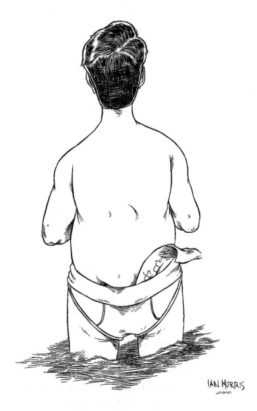

Contemplating a legacy. Image credit: Ian Morris.

REFERENCES

1. Z. Epstein, Horrifying chart reveals how much time we spend staring at screens each day. BGR.com. [newspaper online] 2014 May 29. [cited 2015 October]. Available from: http://bgr.com/2014/05/29/smartphone-computer-usage-study-chart/.

2. H. Kawamoto, History of Liquid-crystal displays, *Proc. IEEE*, 2002, **90**, 460–450.

3. H. Wilson, Labour's plan for science Labour Party Annual Conference. [speech]. 1963 Oct 1.

4. British Library. Inside Science 2013. [video file] 2013 March 18. [cited 2015 Oct] Available from: https://youtu.be/AaB902_ds-g?t=34m.

5. G. W. Gray, J. J. Harrison and J. A. Nash, New family of nematic liquid crystals for displays, *Electron. Lett.*, 1973, **9**(6), 130–131.

6. From archives: The rise and fall of John Stonehouse. Birmingham Mail. [Internet newspaper]. 2012 May 11. [cited 2015 Oct] Available from: http://www.birminghammail.co.uk/news/local-news/from-the-archives-the-rise-and-fall-of-john-stonehouse-153826.

7. Sharp. History of TV making for over 50 years and LCD for over 30 years. [Internet document] 2004 Oct 15. [cited 2015 Oct] Available from: http://www.sharp-world.com/products/lcd_monitor/sharp_lcd/evolution_history/.

Subject Index